Practical risk management in the construction industry

D0881783

ENGINEERING MANAGEMENT

ENGINEERING MANAGEMENT

Practical risk management in the construction industry

Leslie Edwards, BSc, FICE, FCIArb, FIRM

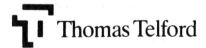

Published by Thomas Telford Publications, Thomas Telford Services Ltd, 1 Heron Quay, London E14 4JD

First published 1995

Distributors for Thomas Telford books are
USA: American Society of Civil Engineers, Publications Sales Department, 345 East 47th Street, New York, NY 10017-2398
Japan: Maruzen Co. Ltd, Book Department, 3–10 Nihonbashi 2-chome, Chuo-ku, Tokyo 103
Australia: DA Books and Journals, 648 Whitehorse Road, Mitcham 3132, Victoria

A catalogue record for this book is available from the British Library
ISBN 978-0-7277-2064-1

Typeset by MHL Typesetting Ltd, Coventry

Preface

This book is aimed at providing initial practical assistance to the student, graduate or qualified engineer who might wish to study, commission, undertake, check or advise on construction risk management matters.

There are now many text books and articles on risk management: some are philosophical or academic, with little practical civil engineering content; some are highly detailed but often focused on specialized applications, e.g. studies applicable to safety in petrochemical plants or weighted towards a particular area of construction risk such as project management. There appears to be little available to readily assist the engineer who wants to assimilate rapidly what the subject is about generally and what others in the industry actually produce as risk assessments and undertake as risk management.

This book, therefore, attempts to provide an easily understood overview of the risk management procedures which are applicable to general commercial organizations, risks that might arise particularly in construction, and then by practical examples and discussion how those risks might be managed.

No apologies are given for excluding philosophical issues such as group decision theory or human attitudes to risk, or for not selecting and discussing in detail individual specialist subjects such as health, safety, environment, fire, or security or other areas that can give rise to organizational risks. It will be apparent on reading the text that, not least because of space limitations, such considerations are inappropriate. The reader should refer instead to authoritative texts on such specialized subjects, many of which contain comprehensive reference lists and bibliographies to assist those wishing to undertake detailed research. The libraries of the Institution of Civil Engineers or the Institute of Risk Management can provide details of these and other risk-related publications.

Having been provided with an understanding of where construction risks sit within the overall commercial risk context, and some practical examples, it is hoped that the reader will be able to ascertain more readily which aspects of risk management or assessment are most appropriate for his or her own purposes.

In the context of this book significant civil engineering projects are envisaged, but many considerations will apply equally to building or other construction. For small structures or parts of projects, similar but more simplistic approaches will often be appropriate.

A short list of references at the end of the book provides a guide to more information on the matters introduced in the text.

Acknowledgements

I would like to thank those various, generally anonymous, technical assessors who have constructively commented during the preparation of this book. Their advice and alternative views were invaluable in encouraging a more critical view to be taken of certain areas of text. I would also like to acknowledge the unfailingly patient secretarial assistance provided by Anne Andrews, Linda Bevins and Sally Feltham who undertook the unenviable task of translating my initial rough notes and numerous amended drafts into something sensible.

Leslie Edwards

Contents

Appendices

References

1 Introduction

Hazardous events can befall any organization and have an adverse effect on that organization's financial well-being. The hazard may be physical, such as fire or theft, which would affect an organization directly, occupational injury or illness to employees which could result in reduced output, injury or damage to third parties or third party property giving rise to a claim for compensatory damages, or a fine due to non-compliance with statutory regulations.

Most organizations operate some form of control over hazards, possibly a formal system, but often one that just happens, e.g. many are content to rely solely upon limited controls by insurances. Such insurances are usually left to the organization's insurance specialist, often with little or no interface with those who are actually undertaking the insured activities. There may be one director with overall risk responsibility, often as a minor subsidiary function to his or her main task, who possibly understands no more than one aspect of hazard management and thus usually undertakes a reactive rather than proactive risk management function.

In the early 1990s the construction industry became more generally aware of a management approach to dealing with hazards, already assimilated by certain other areas of commerce and industry, called risk management. The basic processes are actually quite simple: hazards are identified; the consequences and probabilities of occurrence are assessed; priorities established; the resulting risks are eliminated or reduced and then provision is made for residual risks.

To those who have studied the subject and have had increasingly to undertake risk-related tasks, it has become apparent that risk management can be an effective formalized system with which to address and manage a whole range of hazardous activities to which an organization can be subjected. Far from being yet another non-

productive management overhead, risk management can be used highly successfully to plan ahead in order to reduce adverse effects on company profitability.
Effective risk management provides

- an increased awareness of the consequences of risks
- a focus for a more structured approach to risk management
- more effective centralized management control
- better risk information transfer between those concerned with and those responsible for such matters and, most importantly,
- reduced long-term loss expenditure and hence corresponding increased profits.

It will become clear that risk management is merely a mirror image of general commercial management: the latter is concerned with the identification and implementation of profit making opportunities whilst the former is concerned with reducing the opportunities for, and consequences of, loss. Thus, a longer term view is that risk management should be subsumed within, and be an integral part of, the general management process of all efficiently run organizations, with all line managers having identified responsibilities for the risks in the areas in which they operate. However, for most organizations that is still some way into the future and risk is presently being identified as a growing separate specialized management area. This is endorsed by many leading UK commercial organizations and public sector bodies which now employ a risk manager and, if large enough, have a specialized risk management department. For present purposes, therefore, risk management is treated as an entirely separate aspect of normal business management.

Part 1 of this text describes risk management as it can apply to most organizations, including those inside and outside construction. Part 2 describes other practical applications of the various risk management processes for the construction industry. Appendices are provided which give a brief overview of various common hazard identification and analysis techniques, some of which can be applied directly, or some adapted to apply, to the construction environment.

Part 1
Principles of risk management

2 Hazard, risk, risk analysis, risk management

In any book on risk management it is necessary to understand what is meant by risk management, risk analysis, hazard and risk. There are many different published definitions.

For the purposes of this book, risk management is taken to be

> The identification, measurement and control at most economic cost of the hazards which can threaten life, property and the assets and earnings of an organization.

i.e. it is 'pure' risk management which deals with the concept of loss. There are wider aspects of risk management which can also deal with the concept of gain, e.g. speculative business risks and the activities of portfolio managers who try to optimize investment income. However, for present purposes the pure risk management definition is used.

The words 'hazard' and 'risk' are often used interchangeably. Strictly speaking a hazard is usually considered to be something that might go wrong with adverse consequences, whereas a risk is the multiple of the cost of that hazardous consequence and its possibility of occurrence. In other words a hazard of likely maximum adverse consequence of £100 000 but with a 1 in 10 probability of occurrence is a £10 000 risk.

Risk analysis is the identification and assessment of the likelihood of hazards occurring and the consequences of occurrence. Sometimes the identification of appropriate alternative ways of eliminating or reducing the risk or reducing its impact is included within the definition. Risk analysis is, therefore, a significant initial part of the risk management process.

Commercial basis of risk management

The foregoing definition of risk management should indicate that it has a strong commercial basis. Like any other commercial

4

activity, it must normally compete for organizational funds in the same way as other demands for those funds, e.g. the purchase of new equipment or machinery. There will almost always be a need to show that there is a potential better net gain to an organization in undertaking proposed risk management procedures, compared with the net gain of using the money on something else or investing it directly elsewhere.

This may not mean that the objective is always to obtain the best short-term commercial rate of return. For example, it may only be in the longer term that the creation by risk management expenditure of a public perception that an organization is safe, economically aware, or pollution conscious will result in increased sales or share price. Alternatively, an organization might have a strong commitment to health and safety of employees well beyond its statutory obligations and which cannot be shown to be justified in monetary terms alone. Thus, risk management is always a servant of corporate policy.

Who and what is at risk

At risk from business activities is the organization itself (e.g. loss or damage to property, financial status, intangibles such as reputation and goodwill), directors and officers (negligent activities, non-compliance with statute), employees, third parties and third party property.

Types of risks

Risks can be

- *physical/material* — loss due to fire, corrosion, explosion, structural defect, war
- *consequential* — loss of profits following fire, following theft
- *social* — changes in public opinion, expectations of work force, greater awareness of moral issues (e.g. environment)
- *legal liabilities* — tortious liabilities, statutory liabilities, contractual liabilities
- *political* — governmental intervention, sanctions, acts of foreign governments, inflationary/deflationary policies, export/import restrictions, trading alliances, changes in legislation
- *financial* — inadequate inflation forecasts, incorrect marketing decisions, credit policies

5

- *technical* — increased technology in manufacture, communications, data handling, interdependency of manufacturers, methods of storage, stock control and distribution.

Origin of risks
Risks which might threaten an organization may be those

- *arising from outside the company* — natural hazards, activities of suppliers, debtor customers
- *existing within the company* — physical damage, accidents
- *transmitted from the company* — environmental damage, injury from products, negligence.

Cost of risks
The cost of risks to an organization, whether managed or not, can have a significant impact on its balance sheet.

The cost of risk management itself results from the costs incurred by the identification and evaluation of risks, control measures that might be put in place (such as better security provisions, stand-by plant), the costs of insurance or other financing provisions, and the fees for any outside consultants.

These definite costs must be weighed against the costs if hazards occur, e.g.

- *direct costs of loss* — repairs or replacement of damaged goods or property, third party compensation
- *measurable consequential costs of loss* — loss of, or reduced output, knock-on effect on production chain, losses whilst retraining replacement staff or becoming familiar with replacement equipment, accident investigation costs, lost management time involved in litigation, increased premiums
- *indirect costs of loss* — inability to meet contracts, loss of market share, loss of goodwill, poor industrial relations, poor workplace morale, recruitment problems, poor neighbourhood relations, adverse press relationship.

This assessment of the costs of risk management procedures (including an option to do nothing) compared with the direct, consequential and indirect costs of risks occurring is an essential element in deciding which risk management decision provides the most economic option for the organization.

3 General commercial risks

Many of the hazards associated with commercial undertakings (remembering from the definitions given earlier that hazards in the context of this book are non-speculative in nature and associated with the concept of loss only) are obvious: equally obvious may be possible ways of eliminating, reducing or transferring them. For example, one solution is to insure as many as possible. As will be seen later, this may not always be the cheapest or most effective way of managing risk.

Identifying hazards is an essential part of a structured approach to risk management. The following list identifies many of the hazards in the corporate risk environment. The list (as well as similar ones given later) may be considered somewhat tedious. However, it does give an effective illustration of the wide range of hazards that might need to be managed, as well as providing a convenient checklist to which the reader may add other items in the light of his or her own experience and further reading.

- Liability exposures for death, injury or damage. Liabilities (which will be described in more detail in the next chapter) are those relating to death or injury to persons or to damage to the property of others, caused by an organization's activities. Liability can be to
  - employees
  - third parties
  - consumers for products supplied directly or as component parts to other manufacturers
  - the public at large, e.g. by causing pollution

  There is also liability to the state, for non-compliance with

7

statutory obligations, e.g. Health and Safety, incurring damages, fines and imprisonment

- Building and other property loss and damage hazards include
  - fire
  - theft
  - explosion
  - lightning
  - impact
  - windstorm
  - snow, rain, hail
  - flood inundation
  - frost, ice
  - earthquake, landslide
  - subsidence
  - deliberate damage, sabotage
  - strike, riot
  - war
  - nuclear irradiation or contamination
  - damage during transportation, including marine and aviation
  - money losses
  - glass damage
  - engineering risks (boilers, engines, etc.)
  - loss of profits following property damage
  - latent defects
  - damage from faulty products, in-house or bought in, e.g. faulty material, faulty design or manufacturing, or from faulty machinery

- Motor vehicle hazards include
  - those which might cause damage to the vehicles themselves
  - those connected with death, injury or damage to third parties and third party property

- Criminal risks
  - fidelity (financial losses due to untrustworthy employee)
  - terrorism
  - malicious damage
  - malicious contamination (costs of recalling products, loss

of market share and drop in share price as a result of bad publicity)
- ℗ extortion (threat of action unless money is paid)
- o kidnap and ransom (effect on production and share price of loss of key personnel)

- Hazards dependent on others
  - o poor or non performance by suppliers including internal suppliers of goods and services
  - o poor or non performance by subcontractors (nominated or otherwise)
  - o poor or non performance by joint venture partners
  - o poor quality products and/or not meeting contractual requirements, incurring penalties, cancellations

- For overseas locations or dependency on overseas suppliers or customers
  - o volatile current exchange rates
  - o onerous exchange controls
  - o political climate risks (nationalization, imposed local partners, percentage local nationals in workforce, management)
  - o instability, insurrection, riot, war, terrorism
  - o different statutory obligations and penalties
  - o local insurance requirements

- Financial hazards
  - o inadequate assessment of viability of new products, new plant to be incorporated into works by contractor
  - o tender miscalculations
  - o inflation
  - o statutory pay rises
  - o credit (poor payment of debts)
  - o loss of rental income

- Labour hazards
  - o strikes
  - o working to rule

- Internal management hazards
  - o poor recruitment, job allocation, supervision, training

9

- o inappropriate machinery, guarding, maintenance, usage
- o dangerous materials
- o poor stock control
- o poor production line
- o unauthorized use of resources
- o substandard quality of product
- o duplications, bottlenecks and dependencies in the production process
- o mismanagement of vehicle fleet
- o lack of computer hardware, software and data security backup
- o lack of contingency planning and disaster recovery measures
- o negligent design or advice
- o staff health and safety

- Other hazards
  - o loss of public goodwill, reputation, image
  - o loss of key staff
  - o loss of intellectual property
  - o goods supplied to others becoming property of receiver before goods paid for

- Hazards transferred from others
  - o by contract or by requirement to provide guarantees or warranties
  - o joint and several liabilities where one party can inherit the risks of others more liable

- Actions of competitors
  - o territorial expansionist activities aimed at achieving greater market share
  - o new directly competitive market lines of cheaper, more effective, or perceived more modern products which are more attractive to customers.

As a brief aside, the reader will recognize the extensive efforts and human impact of attempts made by corporate management to streamline costs, increase sales, or reduce losses in order to maintain, let alone increase, market share and profit margins. It is somewhat sobering to realise that a hint of impurity in bottled water, a threat of product contamination in retail outlets, an implication of association with an out-of-favour country or a delay in gaining regulatory approval of a new medical product, might each have a far greater and negative impact on share price and overturn all previous management (including risk management) efforts in the space of a few days.

4 Liability exposures

General

Liability exposures (e.g. the consequences of causing death or injury, or damage to others) are different from most exposures as maximum liability is usually not ascertainable in advance, the fact of legal liability and the cost consequences frequently being determined by the courts, or at least by the threat of court action. Property on the other hand has an upper limit of risk, e.g. the cost of replacing a building. It is for this reason that upper limits of liability are almost always covered by insurance, when available.

Different liability laws apply in different countries. If an organization is operating, or considering operating outside the UK, the liability positions should be determined, e.g. in Eastern Europe there is usually a presumption of fault on the part of a person carrying out an act which results in damage. The impact of EU regulations must also be assessed, both for the present situation and also for possible implications for the future.

Liability in the UK[1] usually arises from common law (tort), contracts between parties, from statute (legislation) and from what is known as strict liability.

Tortious liability

Tortious liability usually arises in the UK from two main groups of causes

- Negligence, nuisance, trespass to persons, goods or land. The former is probably the major cause of concern
- Vicarious liability for
  - o the acts of employees
  - o certain of the acts of independent contractors, e.g. withdrawal of support from neighbour's land, escape

12

of dangerous substances, those contractors negligently selected
o the authorized actions of agents
o the actions of partners.

Liability may be severally (solely to the extent of a party's own liability), or jointly and severally (joint shared responsibility so that a plaintiff can sue one party for the full amount, leaving that party to seek its own recompense from whatever other parties might be involved).

Statutory liability
There is a need to comply with statute, e.g.

- Health and Safety at Work etc. Act (criminal liability for non-compliance leading to fines or a prison sentence)
- Sale of Goods Act (liability for death or personal injury or damage to other property caused by goods)
- Supply of Goods and Services Act (need to exercise reasonable skill and care)
- Occupiers Liability Act (duty to protect lawful and unlawful visitors)
- Control of Pollution Acts (regulations for water, atmospheric noise, land pollution and for the disposal of waste materials)
- Companies Act (directors and officers liability for wrongful acts which include neglect, error, breach of duty, failure to insure, failure to supervise, failure to comply with the requirements of statute).

Strict liability
Strict liability arises where an organization is deemed liable in law for damage, even though there was no intent to cause damage nor was there negligence. Examples, which are part of case law, include

- liability for damage caused by the escape of something not naturally present on land under the control of the party being sued
- escape of fire.

Statute can also impose strict liability, e.g. it is a strict liability to comply with the Health and Safety at Work etc. Act.

Contractual liability

Wide ranging liabilities can be included within contracts between parties, and contracts may need careful study to determine what these are. Once signed, it is extremely difficult to avoid the consequences of non-compliance with the requirements of a commercial contract freely entered into.

Consequences

Liability can give rise to

- damages ranging from nominal to aggravated, but usually to put the injured person in the same position as before damage occurred
- compensation for death, or for personal injury. This will cover past and present loss of earnings plus a payment for pain and suffering. (Note that for several persons injured by one negligent act the compensation can be large in total)
- fines and/or imprisonment for statutory breach, e.g. for breach of health and safety provisions. From 1992, these can be £20 000 in Magistrates Courts, or an unlimited fine in the Crown Courts plus two years in prison.

5 Risk management

Risk management involves risk analysis, control, transfer and financing. Figure 1 gives an overview of these processes and forms

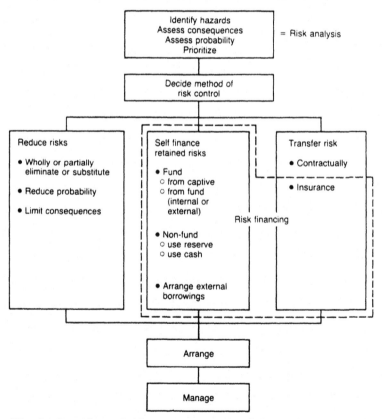

Fig. 1. Overview of risk management practices

15

the basis for much of what follows in the text. An organization's risk management programme will usually involve a combination of the principal elements.

Periodic review of an organization's exposures and its risk management programme mix is needed because, for instance, exposures and insurance/reinsurance/financial market conditions change. In addition, as always necessary for good management, the effectiveness of the programme should be audited, whether or not the exposures or risk management mix are altered.

6 Risk analysis

General

Risk analysis comprises the following activities

- identification of hazards
- quantifying the financial consequences if the hazard occurs
- determination of the probability of occurrence
- calculating the risk and prioritizing in order of importance.

Hazard identification

As many hazards as possible which might pose a risk to the aims of an organization, a part of an organization, or a project, need to be identified. The aims are usually commercial, e.g. the maximization or maintenance of profit margins, dividends, share price. For other organizations, e.g. local authorities, health authorities, it may be threats to the continued provision of services to the public, and for others (say, in the case of water companies) it may be a statutory obligation to maintain quality. Identified hazards should have had proper risk assessments and adequate provisions made. They should not, therefore, have a long-term impact on the achievement of organizational aims. It is often unidentified hazards, for which there is no provision, that have the most significant impact. Perhaps it is for those unidentified hazards that true contingency provisions should be made.

Examples of methods for identifying hazards are

- analysis of available records
- brainstorming sessions
- physical inspections
- check lists
- organizational charts
- flow charts
- hazard and operability (HAZOP) studies
- fault trees, example of hazard analysis (HAZAN) process
- hazard indices, e.g. Dow Fire and Explosion Index.

Some of those methods are described in further detail in Appendix 1. A method appropriate to the relevant circumstances should obviously be used, e.g. for construction the most common method will often be some form of committee brainstorming session.

Assessment of consequences
The consequences, usually financial, of a hazard occurring must be determined. The cost of risks have been described earlier — these must be quantified. For example, effects on a production process, both 'upstream' and 'downstream' of a hazard event, must be considered. Some methods of identifying hazards include structured ways of assessing those upstream and downstream consequences. Also to be taken into account under this heading is the number of exposure events over the period concerned, i.e. a large number of small losses can have a similar effect on a balance sheet to those resulting from a single loss. It may not therefore be valid to ignore individual events whose impact might otherwise be considered to be small.

Assessment of probability
Probability of hazards occurring may be determined from

- historical records of the organization, or the organization's insurers
- industry-wide records
- site specific records of physical risks such as the return periods of windstorm, earthquake, rainfall, flooding, etc.
- manufacturer's data relating to plant planned outages, e.g. for maintenance
- subjective assessments
- mathematical analysis of loss data.

Some mathematical aspects are discussed briefly in Appendix 2. However, sufficient data may not be available in many construction circumstances to make such analyses very meaningful.

Determination of risks and establishment of priorities
Once hazards have been identified and their potential loss creating capacity and likelihood of occurrence have been assessed for any one period (typically a financial year), an order of priority can be established.

This is usually done by using the formula

Risk = Potential Loss × Probability of Occurrence

Table 1 gives a very simple example of how four risks of known consequence and probability can be prioritized using the foregoing risk formula.

Table 1. Use of the risk formula to determine hazard priority

Hazard no.	Loss potential	Probability	Risk	Ranking
1	£1m	1 in 100	£10 000	3
2	£0·5m	1 in 20	£25 000	1
3	£1·5m	1 in 200	£7500	4
4	£0·20m	1 in 10	£20 000	2

The present text deliberately excludes more academic or philosophical considerations such as the effect of human behaviour, group decision-making theories, etc. to concentrate on practicalities. However, it is worth digressing briefly at this point. The reader might use the risk formula to ascertain what is considered to be the most significant risk: a 1 in 100 chance of a £1000 loss or a 1 in 100 000 chance of a £1m loss? Clearly the formula gives the same result, but the reader might have a firm opinion that one risk is more significant than the other. Alternatively, what about a 1 in 1000 chance of a single human injury or a 1 in 1 million chance of 1000 injuries? The reader has thus strayed into the nebulous area of risk perception, indicating that it is not just in pure subjective assessments where the risk inclinations of individuals or groups may come into play.

Subjective assessments
Sometimes it is not possible to ascertain quantitatively the factors which should be applied to identified hazards in order to progress a risk assessment. Subjective judgement is often possible — indeed it may be the only way forward.

Subjective severity grading
One way of assessing severity is in terms of the financial impact on an organization, as shown in Table 2.

Table 2. Severity grading[2]

Grading	Subjective assessment of the severity in general terms	Subjective assessment of the severity in money terms
1	Nuisance, current expense impact only	*
2	Medium losses within the margin of the insurance deductible (or excess)	*
3	Manageable losses	*
4	Range of largest previous losses	*
5	Serious losses	*
6	Most serious Catastrophic — total loss type	*

* Severity descriptions should be defined in relation to the organization concerned. This may lead to a decision to increase the grading range for more sensitivity.

Subjective probability assessment

A relative probability can also be derived by subjective judgement, e.g. on a scale of 0 to 1·0. Examples of probability scales are given in Tables 3 and 4.

Table 3. Probability scale

Probability factor	What the factor signifies
0	Loss is not possible
0·1	Possibility is very remote
0·2	Remotely possible
0·3	Slight chance of it happening
0·4	A little less than equal chance
0·5	Equal chance of it happening
0·6	Fairly possible
0·7	More than likely to happen
0·8	Predictable
0·9	Very probable it will happen
1·0	Loss is certain

Table 4. Probability scale

Probability factor	What the factor signifies
0%	Nil chance
5% to 45%	Unlikely
45% to 55%	As likely as not
55% to 95%	Likely
95% to 99%	Almost certain
100%	Certain

The application of the scale in Table 3 is illustrated in Table 5.

Table 5. Subjective assessment of probability on a scale of 0 to 1

Hazard no.	Loss potential	Subjective relative probability (0 to 1)	Risk	Ranking
1	£1m	0·2	£0·20m	2
2	£0·5m	0·5	£0·25m	1
3	£1·5m	0·1	£0·15m	4
4	£0·20m	0·8	£0·16m	3

The answer is still not particularly scientific and there is clearly scope for manipulation. The basic principles are, however, sound and give a management rationale for decision making.

7 Risk control

General

Once risks have been identified in order of priority, it is usually possible to see that some are unlikely to affect noticeably the organization's balance sheet, whereas others can be highly significant. However, as already noted, it may not be valid to reject entirely risks of low value if there are likely to be many of them. Clearly, many low value risks can be as significant as a single larger one.

Each significant risk must be considered in terms of

- elimination of the hazard, e.g. no hot cutting/welding on premises
- substitution of the hazardous substance or activity by a less hazardous one, e.g. use more prefabricated structural components for construction over water
- reduction of the hazard, e.g. more guarding/watching to reduce theft
- reduction of the probability of occurrence, e.g. training and education of personnel in use of specialized equipment
- minimization of the consequences, e.g. the provision of back-up and standby equipment, emergency planning procedures
- contractual transfer or financial provision for residual risks.

Whichever method or combination of methods is used, the end result should produce the most economical solution to the management of risk.

Residual risks

Risks which cannot be totally eliminated, substituted or contractually transferred to others are residual risks.

The costs consequences of some residual risks occurring, particu-

22

larly those of a more random nature, can be spread uniformly over a number of years as payment of premiums to an insurance company.

The costs of the remaining self-retained risks occurring must be met by alternative ways, usually described, perhaps somewhat erroneously, as risk financing. For convenience, insurance is also usually included within risk financing.

In the following chapters, contractual transfer and alternative methods of risk financing are described.

8 Contractual transfer of risk

Theory of risk allocation

The responsibility for indemnifying the consequences of a risk event resulting from the activities of one of the contracting parties should ideally rest with the party who has control over that risk, e.g.

- If the actions of client's staff, negligent or otherwise, result in damage to works being undertaken by a contractor, then that should be a risk indemnified by the client.
- If a contractor's employees or equipment damage a client's property, then those costs should be borne by the contractor.

In practice, it is usually best commercial policy that responsibility for such risks should rest with the party best able to manage them, e.g. the party with the relevant insurance cover. The actual sharing of risk, indemnities and provisions for supporting insurances will be determined by the wording of the relevant contract documents.

For other areas of risk not caused by the actions of either party, standard forms of contract usually share the risk between them. For instance, in the case of weather, the risk is often apportioned according to whether the weather condition is exceptional or otherwise, it usually being deemed uneconomic to a client for contractors to accept and include within their tender prices for the risks associated with the exceptional weather conditions.

In fact, many contracts are non-standard, or standard forms are made non-standard by additional clauses. The consequence is usually to attempt to transfer more risks from the drafter to the tenderer. The practicalities of this are limited by certain basic requirements which, if not met, probably make the transfer pointless. There are also statutory restrictions.

Basic requirements for risk transfer

The requirements for those to whom risk is being transferred include

- ability to undertake a hazardous task
- willingness to take the risk
- financial capability if the risk event occurs
- continued existence and adequate finance during period of liability.

Statutory limitations on contractual transfer

The Unfair Contract Terms Act 1977 prohibits the exclusion or restriction of liability for death or personal injury resulting from negligence. Exclusion of liability for other loss or damage is subject to 'a test of reasonableness'.

The Act also makes it impossible to contract out of certain other statutory liabilities such as those relating to the sale and supply of goods for private use or consumption.

The Consumer Protection Act 1987 and its EU equivalent imposes no-fault liability on 'producers' for death, personal injury or property damage to consumers. Producers are widely defined and can mean almost anyone on the chain of manufacture or supply. The result is that any producer, even a retail outlet, can be sued in isolation for faults that occurred somewhere else in the chain. It is therefore essential for all producers to have indemnity clauses, or similar, in place with others in the chain who may be more properly liable.

Transfer of risky activity

Risky activities can be transferred by

- *contracts, subcontracts*, e.g. having risky work undertaken by others. Residual vicarious liability by a principal for certain actions of a subcontractor may remain, e.g. the removal of support from adjacent land, the escape of fire or dangerous substances.
- *property, vehicle, machinery leases*, e.g. the transfer of the repair/maintenance responsibility.

Transfer of financial consequences of risks
The financial consequences of risks occurring can be wholly or partly transferred by means of

- *indemnities* — agreements to pay costs of losses to property, damages for liability
- *'hold harmless' agreements* — types of indemnity dealing with legal liability claims
- *sureties* — agreements by a third party within the framework of the main contract between the two parties to pay money in the event of non-performance by one of those main parties
- *bonds* — agreements to pay money if quality or fitness for purpose are not met
- *guarantees* — agreements to provide recompense for inadequate products or services. This is a separate contract wholly outside the main contract
- *insurances*
- *liquidated damages* — agreement to provide recompense for the effects of delay.

Resistance to risk transfer
Contractual transfer of risks is all very well to the party doing the transferring. However, the receiving party may not want to accept those risks because of its own risk management controls. Various courses of action available to the receiving party can include

- a refusal to accept certain proposed clauses
- a refusal to provide retrospective collateral warranties, guarantees etc.
- an insistence on additional specific exemption clauses to avoid certain exceptional risks, obligations or the possible consequences.

9 Risk financing

A definition of risk financing is

> To ensure the economic provision of funds to finance recovery of an
> organization from property damage, liability claims from employees
> and third parties, personal injury/death affecting the efficient running
> of business, and business interruption losses (normally loss of profits
> for a pre-agreed limited period).

The term 'economic provision' means the most economic choice
of alternative form, or combination of alternative forms, of providing finance to meet the consequences of adverse fortuitous events
affecting the organization.

Methods of risk financing

Risk financing can be by one or a combination of the following,
each of which will be described in more detail in following chapters

- insurance
- captive insurance
- internal/external funds
- contingency reserves
- cash borrowings from income
- external borrowings
- financial reinsurance.

Some small organizations might comply with statutory insurance
requirements only and rely on taking money from cash flow or
loans to finance residual losses. Multinational organizations might
have highly complex arrangements incorporating many of the
foregoing options with different insurance companies in different
territories covering different exposures, legal requirements, and
using more than one international insurance broker to manage the
multinational's insurance portfolio.

10 Insurance

Nature of insurance

Insurance is a mechanism for smoothing the costs of losses, not a loss transfer mechanism, i.e. in the long-term losses have to be paid for out of premiums which also have to pay for insurers' overheads and profits. It is used, therefore, to assist guarantee at a regular annual cost the future financial stability of an organization against the consequences of irregular non-speculative business risks. For insurances, other than those required by statute (see later) or by contractual agreement, it is sometimes described as a 'sleep soundly' option favoured by those who do not have the size of exposure to loss, staffing capability, knowledge, advice, or inclination that might warrant a more sophisticated approach.

Insurance should be used intelligently. For example, if annually from records an organization regularly incurs losses of around £100 000 in global company motor vehicle claims, there is little point in paying an insurance company £135 000 (a figure including a possible 35% insurance company overhead and profit mark-up) of insurance premium to cover such risks — this is known as pound-swapping. The £135 000 is best kept within the organization to be used for self-financing, earning interest at the same time. Formal insurance is better used, and is often the cheapest alternative, for more unpredictable and particularly for catastrophic property or liability losses (i.e. those that might affect the continued viability of an organization). It is therefore best used for smoothing out the consequences of peaks in losses in excess of a reasonably regular annual norm.

There are several significant disadvantages perceived for insurance.

- Certain risks may be only partly insurable, e.g. it may be difficult to find insurers for riskier areas such as terrorism, gradual pollution or other long-term liabilities, earthquake cover in certain territories, and so on.

- Premiums are often based on the claims experience of a pool of similar insureds. In such circumstances, premiums do not reflect good risk control or good claims history of a particular client.
- Insurance has a high mark up covering overheads and profit. It has been described as the equivalent to putting £100 into the bank but only getting £65 back.
- The cover available and premiums charged can be very variable, and are often subject to renegotiation on an annual basis. This can be to the detriment of an insured who may not be in a position to organize a better alternative. It will also be noted that the greater the variation in terms and conditions the less an insured can rely on insurance to contribute to future financial stability. This leads to consideration of other options.
- The service offered by many insurers is perceived to be poor, e.g. there are claims disputes or delayed payments.
- There are usually bottom end deductibles and top end ceilings on cover for which alternative provisions may need to be made.
- There is no protection against criminal liability, e.g. no compensation for fines and prison sentences.

Because of the disadvantages listed, accepted risk management theory is that insurance should only be used as a last resort. It has been quoted as "a complicated and inefficient method of borrowing money, the essence of which is that you pay back the loan before you get the money!"

In practice, for the reasons given at the start of this section, and because of statutory and contractual requirements for insurance, direct insurance is a very common method by which most small to medium organizations manage risk exposures. In addition, for large organizations, insurance is often the cheapest method (when available) for managing catastrophic-type physical risks, e.g. terrorism, storm, earthquake, and for managing the higher levels of liability exposures.

Insurances which might form part of insurance programme
Liability insurances
Because liability is open ended, i.e. maximum liability cannot be accurately predicted in advance, it is usual for liability exposures

to be covered as far as possible by some form of insurance. Legal liability is usually for bodily injury or death to persons, or for damage to third party property. Insurance cover is limited in many countries, e.g. North American liability awards can be far above what insurers are prepared to cover fully. Liability insurances include

- *Product liability* for liabilities to users. Also product recall insurance (accidental defects), product guarantees, consequential loss insurance.
- *Employers liability* (EL) insurance for liabilities to employees. This is compulsory in the United Kingdom.
- *Public liability* (PL) insurance for liabilities to the public.
- *Professional indemnity* insurance for liabilities relating to the provision of services, including design and advice. It covers negligent acts, errors or omissions.
- *Directors and officers* insurance for liabilities of principals resulting, for instance, from negligent acts during the course of company activities.
- *Motor* insurance. Compulsory in the UK for the third party legal liability element of cover.

Property insurances
Including

- general all-risks (which include fire, theft, flooding, etc.)
- business interruption
- goods in transit
- money
- glass
- engineering (boilers, engines).

Motor vehicle insurance

Marine and aviation insurance

Miscellaneous insurances

Including

- credit (poor payment of debts)
- fidelity (losses due to untrustworthy employees)
- loss of rent

- terrorism
- malicious contamination
- extortion
- kidnap and ransom.

Contractors all risks, erection all risks (usually in conjunction with public liability) insurances

These are looked at in more detail later in this text.

Development of risk management as a consequence of insurance disadvantages

Historically, most organizations did not become particularly interested in risk management until insurance cover either became unavailable or became noticeably more expensive than the organization had hitherto been used to. It is problems with insurance which have led to better awareness of the benefits of improved risk management and the value of all-round trained risk managers. Up until fairly recently, most risk managers in the UK started as an organization's insurance manager with little or no training in risk management outside of insurance, little practical familiarity with the alternative risk financing options available and no incentives to become involved with them.

In the USA, which is traditionally several years ahead of the UK in management practices, risk managers are increasingly likely to be more widely trained and have a core discipline outside insurance. A consequence of more effective risk management is that with larger, better managed organizations partly opting out of the insurance market, insurance companies are being left with poorer risk organizations in their pools and with an increasing number of the more unpredictable and higher risks. Inevitably premiums must increase, terms become more onerous and the spiral continues. This should, of course, result in the more poorly managed organizations or those with more difficult risks paying a fairer portion of their relevant costs and there will be pressure on them to undertake better risk management.

11 Captive insurance companies

Captive insurance

An alternative or addition to using insurers in the direct insurance market is to use a captive company. A captive is a subsidiary insurance company of a non-insurance parent. One or more captives are increasingly becoming part of a solution which might include direct insurance and other means of self-financing (see next chapter) for covering the self-insured portion of a large corporate parent's risks.

A captive can provide a tax efficient method for large organizations to fund otherwise expensive or uninsurable risks. It requires an early substantial capitalization to cover possible large early claims. The captive manages the parent organization's risks. It can retain some risk directly, it can use the direct insurance or reinsurance markets to cover larger risks on its behalf and, less usually due to lack of experience or due to parent organization policy, it can also underwrite outside businesses.

Pre-existing captive companies are available, set up and run by specialist companies for use by others.

Advantages associated with a captive are that

- premiums are tax deductible as for any bona fide insurance company
- they can be set up in offshore locations where the capital gains tax on profits can be cheaper
- premium is kept within the company and hence can be used to earn interest
- premiums charged by a captive to a parent company can reflect the risk control measures and actual claims record of the parent company
- there are no costs of marketing, no profit margins, no costs

of inefficient management, so premiums are less
- access to cheaper reinsurance company markets for the riskier portion of the captive's risks *
- there is prompt payment of claims as a captive is acting in parent company interests.

To obtain the stated advantages, a captive must be able to demonstrate to the UK insurance regulatory authorities that it is, in fact, a genuine insurance company and not solely a tax avoidance device, and must show that it is acting at arms length from its parent organization. It must have a board which meets a number of times a year, a registered office, fulfil various statutory requirements regarding, for instance, its ratio of assets to liabilities, submit audited accounts and returns to the government, fix and collect premiums, write policies, manage and settle claims, etc. To obtain certain significant tax advantages these activities will need to take place outside the UK. It will be evident that all these matters are costly, particularly so when the business is first being set up, with more regular operating costs thereafter. Where certain activities are delegated to outside consultancies, costs will still be incurred. On top of these costs are the risks associated with running an insurance company, i.e. the chances of making a short- or long-term loss, which depends upon the skill of the captive's management and on fortuitous events which can give rise to claims.

It will be evident that a potential expenditure of equivalent direct insurance premium of several hundreds of thousands of pounds is necessary before an organization is likely to make a profit by operating a captive insurance company instead of using the direct insurance market. A decision to become involved in captive insurance is usually only taken after a detailed investigation by specialist brokers or consultants of the parent organization's risk exposures and claims history.

Reinsurance

Reinsurance is a wholesale market for insiders such as direct insurance companies and captives which allows them to spread their risks. A large amount of construction insurance is covered ultimately by reinsurance.

Reinsurance companies are approached through specialist reinsurance brokers. Reinsurance terms and conditions are often negotiated on the basis of the actual historical loss records and risk management/loss control measures of clients. (In the present context a captive would be a 'client'.) Therefore, they favour better risk clients. There are lower costs as reinsurers are generally more specialized and efficient than direct insurance companies and do not have the costs of 'retail outlets'.

A wider range of risk cover is available, i.e. areas of risk where it may be difficult to obtain cover on the direct insurance market. Moreover, a higher level of cover is available. Cover can be more flexible, e.g. underwriting only certain higher layers of risk that direct companies will not cover.

Reinsurance can be used also to reduce the minimum regulatory capitalization requirements of a captive insurance company. This is because in some countries regulations interpret liabilities to be lower where more risks are reinsured and hence the capital assets required by regulations to be held by captives are correspondingly lower. This can clearly be important to a decision as to whether and where to operate a captive.

The reinsurance market has become so important to the UK insurance market in some areas that a decision by reinsurers to restrict level of cover, or a reduced insurable capacity by reinsurers, is directly reflected in the cover available from direct insurers to clients.

12 Other ways of risk financing

Previous chapters have described how risk financing can be used to cover the consequences of residual risks occurring. Insurance has been described as one way of smoothing some of those losses. Captive insurance companies have been discussed as an alternative way for organizations with large direct insurance premium expenditure to invest that money more cost effectively. Where it is not economically viable to set up a captive company or by choice not to have one, or where residual exposure remains even when a captive is in operation, other ways of self-financing the cost consequences of retained risks can be considered.

Internal or external funds
Specific internal fund
An organization internally sets money to one side to be used solely for funding self-retained risks. A potential problem is that there may be pressures to use the fund for alternative purposes.

Pools or mutual funds
These permit similar smaller organizations to combine to carry larger risks. There is a certain loss of confidentiality and care is needed to select partners with a similar long-term loss record and positive attitude to risk control.

Insurer's risk fund (sometimes called an external risk fund)
Money is paid to insurers who set up a fund. Contributions made to insurers are tax deductible. Insurer's liability is only to the balance of the fund. Unfunded portions are usually covered by insurers as part of an overall package of insurances. Premiums are based on, for example, five years of client's claims data, with added insurer fees, inflation allowance, etc. Money in such a fund may be reinsured, i.e. the insurer acts as a captive.

Contingency reserve

A portion of funds is earmarked on paper in the organization's accounts to equate to expected losses. It is an accounting device which has no protection from internal paper transfer to other areas if there is appropriate demand.

Cash borrowings

Pay losses from income

This is most appropriate where losses are reasonably predictable in terms of rate of occurrence and size.

A positive 'do nothing' policy

As described earlier, the cost of risks includes the costs of risk management and risk financing, the direct costs of losses (i.e. repair or replacement), measurable consequential costs and indirect costs. In practice, the part of these costs covered by insurance can often be only a relatively small part of an organization's total potential losses, the remainder of which is often covered by other means without large balance sheet effects. It has been suggested that even the insured costs could paid for more cheaply from cash flow instead. 'Do nothing' is not a popular concept but should always be considered in the context of total expenditure and return. Much will depend upon the likelihood of catastrophic loss or of large liability claims. There is certainly a direct incentive for better risk control measures. 'Do nothing' is different from the 'pay (minor) losses from income' concept because it also could be used to cover larger losses. It is not known whether this procedure has ever been adopted as deliberate policy, but it may be worth considering for some situations.

External borrowings

Pre-loss financing arrangements

The rate of standby credit from banks where total fees may be smaller than from insurers should be considered. It can be made available immediately, usually for a minimum of five years. A commitment fee up front is also required which has short-term balance sheets effects.

Post-loss financing

Overdrafts, term loans. By definition, post-loss terms and

conditions which may be onerous are not known until after a loss occurs.

Long-term finance derived from the sale of stocks and shares or from secured loans take time to organize: they are thus no good for a rapid response to catastrophic losses but may be useful for large non-urgent, non-catastrophic losses.

Financial reinsurance and similar

There are various financial procedures that can be undertaken for hedging risks, particularly catastrophic risks, should they occur. One type are long-term contracts based on initial assessments with post-contract premium adjustments relating to actual claims experienced. This is a specialized area outwith the scope of this present text. Effective use is limited in the USA and UK by accountancy regulations. In the USA, exchange trades futures and options are also sometimes used.

13   Risk management decisions

The risk financing options available to an organization to manage its residual risks have been described previously.

Which option or combination of options will be chosen finally depends upon many issues including the organization's policy, its financial strength, size, the risks it can afford to self-finance, and so on.

Table 6 gives typical relationships between different sorts of organizational risks and commonly valid corresponding risk management solutions.

The actual risk management programme adopted by an organization will be the result of a detailed investigation of the organization's past, current and future risk exposures.

Table 6. Relationship between company risks and risk management solutions[2]

Type of loss	Frequency	Severity	Predictability	Impact	Solution
Trivial	Very high	Very low	Very high	Negligible	Non insurance
Small	High	Low	Reasonable Within 1 year	Insignificant	Self insurance
Medium	Low	Medium	Reasonable Within 10 years	Serious	Part Self insurance Part insurance
Large	Rare	High	Minimal	Catastrophic — affects continuing viability of company	Insure

14   Role of a risk manager

The co-ordination of risk management activities within an organization is increasingly the responsibility of a designated risk manager. It may be a combined role, e.g. insurance and risk manager, or it may be entirely separate.

A primary aim of a risk manager is to advise on the optimization of resources within an organization to assist achieve the most cost effective balance between expenditure on loss reduction and the losses that might occur.

A risk manager should ideally be especially knowledgeable about the main risk areas of concern to his or her organization and know enough about the subsidiary or more general areas to be able to co-ordinate all aspects efficiently, knowing when to seek appropriate advice. For example, a risk manager may be a specialist in law, security, fire protection, health, computer security, engineering, etc.

In practice, the main tasks of a risk manager will almost always be in insurance and alternative risk financing. Risk managers can spend typically 50% or more of their time on pure insurance-related matters.

Responsibility

A risk manager is usually responsible to an organization's main board. Day-to-day communications will be to a board member, usually the finance director.

The risk manager's role at operating level will be essentially advisory, day-to-day management of risk being the responsibilities of line managers following advice and training provided by the risk manager.

The typical position of the risk manager within a large group is illustrated in Fig. 2.

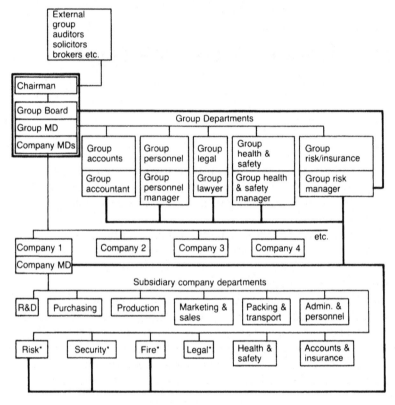

Fig. 2. Group organigram showing typical position of risk manager

Tasks

In practice, the role of a risk manager can vary between organizations. As it is a relatively newly recognized profession in the UK, both risk managers and their organizations often learn as they go along. Indeed, most risk managers are avid readers of how others tackle their roles.

The tasks of the risk manager will depend upon his, her or the organization's experience, the size of the manager's department, the complexity of the organization, and so on. Tasks could include

- assistance with the formulation of an initial risk management policy statement for discussion and approval at board level
- to assist in the agreement of an organizational risk financing strategy
- undertaking initial and then periodic general reviews of current risk management procedures, staff resources and loss financing within the organization
- identification of those matters requiring priority attention including risk management lines of communication
- the provision of advice and assistance to line managers on how to systematically and regularly deal with risk identification, quantification, control and prioritization
- the ongoing identification, recording, quantification and reporting of the physical risks and loss trends associated with the organization's activities
- the ongoing identification and reporting of business, economical, political, litigious, statutory and other risks associated with the countries with which the organization is associated in its purchasing, production and sales operations
- the submission of proposals for the optimum allocation of resources to eliminate, reduce or otherwise control the risk of losses
- co-ordination of a risk financing programme including insurances
- advice and assistance with contingency plans for tackling major losses
- the education and training of employees in risk matters
- the preparation of risk-related reports
- the engagement of external risk consultancy services where appropriate
- liaison with authorities concerned with hazards arising from the organization's activities including employee organization representatives.

It will be clear that significant experience, or at the very least wide ranging and relevant management training, maturity and communication skills, is required for such a role.

15   Training

Presently, risk managers of leading UK organizations are mainly recruited from insurance or from broker's offices, where risk experience, often specialized, has been gained as a relatively specialized part of professional insurance examinations and narrowly focused on the job training. They are now increasingly members of the Institute of Risk Management which was established in the UK in 1986. There are presently (1995) approximately 1000 members of whom about a quarter are Fellows.

The Institute's aims are similar to other professional societies such as the Institution of Civil Engineers, e.g. they include to determine required standards of expertise, hold examinations, grant certificates, promote the profession and to encourage technical and ethical good practice.

Existing Fellowship of the Institute is dominated by experienced risk practitioners from industry, invited to join when the Institute was first established. Increasingly Fellowship is now being gained by private study examinations involving a selection from the following subjects

- risk and the business environment
- business finance
- risk analysis
- physical control of risk
- insurance

- liability exposures
- risk financing
- occupational safety and health
- corporate risk management.

An alternative to private study is to obtain a degree in risk management. At present there is one tertiary educational establishment in the UK offering a totally dedicated qualification.

Currently candidates for Fellowship must also have five years relevant experience and produce an original dissertation on a risk management topic. There is compulsory continuing professional development (CPD) following election.

Part 2
Construction risk management

16 Construction hazards

Part 1 of this text described the hazards associated with running any commercial business, e.g. property, liability, personnel and business interruption risks and ways of managing those hazards.

Part 2 now deals with additional risks associated with the construction industry, hazards associated with particular projects with which parties are connected and ways in which the earlier techniques can be used to manage them.

Consequences

Construction hazards occurring can result in

- failure to keep within budget
- failure to keep within the time contracted for handing over
- failure to meet the required technical standards for quality, or fitness for purpose.

The ultimate consequences are essentially financial. The need for risk control and management should therefore be self-evident.

Risk controls

As for other areas of commercial activity, risks or their financial consequences can be

- *eliminated, reduced, controlled.* For instance, design requirements can be altered, construction methods or materials changed, more investigation undertaken, more training given or more supervision provided. As a last resort there may be a refusal to tender for work which is too risky.
- *transferred contractually to others, or insured.* For example, risks can be transferred
    - from client to contractor or insurer

   o from contractor to subcontractor, or insurer
   o from designer to insurer

- *retained* and financial provision made for hazards that might occur by self-financing.

The extent of risk control and management must, as always, be appropriate to the size of risk.

Extent of detail required by analysis

For all but the simplest projects (which presumably then begs the question as to the necessity for any formal risk assessment process anyway), there should be a limit placed on the number of potential risks that are assessed. It is sometimes suggested that it is unnecessary to analyse further than the next risk on a list of risks of decreasing importance, where that risk has a negligible impact compared with the total effect of the risks that have gone before it. This may not necessarily always be correct (e.g. where there are a large number of risks of different ranking but similar weighting), but it appears to be a useful guide which will be valid in many circumstances.

Different perspectives on risk

A common and prime concern of each party involved in a construction project will most usually be to maximize or at least make a reasonable profit or rate of return. Failure to achieve that objective may have different consequences to different parties. Those parties will therefore usually see risks from different perspectives.

The following are not definitive statements and the interested reader might wish to form his or her own opinion on their validity and of possible additions or alternatives.

- A promoter (financier or client) is concerned that whatever happens on a project, a reasonable rate of return is obtained for the risks undertaken. He is concerned, therefore, in a broad brush way with how likely changes in estimated costs, benefits and timing will affect that rate of return. Ultimately, he has to justify decisions and investments to shareholders, otherwise they will ask why money was not invested somewhere safer and more productive.
- A contractor has similar concerns but might see risks as relating

to maintaining short-term cash flows and maximizing long-term profitability. He will be more concerned than a promoter with the detail of the risks associated with a project and with not exceeding his estimated cost for the works.

• A professional consultant will view project risks from the point of view of accurate fee estimates, good management, prompt and full payment of fees due, the minimization of liability exposures for negligence, the costs of uninsured risks and the future costs of insurance premiums.

• Insurers will only be concerned with the risks that they have insured. Their prime concern will be to ensure that they have accurately estimated those risks, imposed the correct conditions (e.g. excesses and ceilings) and charged the correct premiums. It has been suggested that insurers like claims as they are indicative of the scale of insurance business. Insurers need market activity and premium income to justify their existence. As long as they have agreed the correct terms and conditions to achieve or improve on their projected annual turnover and profits, claims are a natural acceptable part of the insurance cycle. If there were no claims there would be no interest in insurance, no turnover and no profits, hence no need for an insurance market. However, in a hard market, premium rates may need to be low to keep market share and insurers may need to rely solely on income from the investment of premium income to make a profit. In those circumstances, insurers are far more interested in reducing claims, not least by becoming more concerned with their, and the contractor's, risk management procedures.

Risk management as required by each of the foregoing parties will now be considered in more detail. The differentiation between funders, clients, contractors and professionals is recognized as somewhat arbitrary as clearly a contractor can be a client to subcontractors, a client can be a provider of capital funding, and so on. However, it is considered that the described breakdown provides a logical framework in which to address different forms of practical construction risk management.

17 Feasibility study risk assessments for funders

The greatest uncertainty on a project is at the feasibility study stage when much less is known. Three common methods of undertaking risk assessments at that time are sensitivity testing, Monte Carlo analyses and MERA (multiple estimating using risk analysis) estimates. Each can be used for initial project evaluation purposes and periodically as part of ongoing project management. The methods, or those similar to them, can either be developed in-house, can be bought in as stand alone computer packages, or are available as add-ons to project management packages. They vary in terms of depth of detail that can be assessed, type of output, convenience and speed of use, need for specialist involvement and so on. Whichever method is selected, it should be appropriate to the size of project and the purposes for which the end results will be used. For instance, a large potential World Bank multi-discipline overseas aid project may require a different degree of assessment than a smaller local project to which a UK developer is already financially committed.

Economic/financial project evaluations
To understand methods of risk assessment it is necessary to understand firstly some of the basic principles of project evaluation. Such economic/financial analyses are usually undertaken initially at the feasibility study stage to convince a funder (such as a lending agency) that investment in a new project is worthwhile. Initial best guess estimates are used for this.

A typical procedure is to

- determine the demand for a project, typically by investigation and forecasting
- cost the different ways of meeting that demand

- compare the total costs of the cheapest alternative with the total costs of the perceived benefits to determine if acceptable financial criteria are met.

In the following simple example (Tables 7 and 8) two alternative

Table 7. Alternative 1. 40 MW gas turbine power station

| Year | Costs (£ thousands) | | | |
	Construction costs	Plant costs	Operating costs Fuel	Total costs
1996	1000	—	—	1000
1997	1000	500	—	1500
1998	1000	500	—	1500
1999	—	—	300	300
2000	—	—	300	300
To	↓	↓	↓	↓
2010	—	—	300	300
2011	—	—	300	300
etc.	—	—	300	300

For discount rate = $x\%$
Net (1996) present value of total costs = £7 000 000

Table 8. Alternative 2. 40 MW hydro-electric power station

| Year | Costs (£ thousands) | | | |
	Construction costs	Plant costs	Operating costs Fuel	Total costs
1996	500	—	Negligible	500
1997	1250	—	—	1250
1998	1500	750	—	2250
1999	750	750	—	1500
2000	—	—	—	—
To	↓	↓	↓	↓
2010	—	—	—	—
2011	—	—	—	—
etc.	—	—	—	—

For discount rate = $x\%$
Net (1996) present value of total costs = £6 000 000

means of providing 40 MW of power are illustrated and costed. It is postulated that for a specified discount rate of $x\%$, a net present value for the 'Total costs' column relating to a datum year of, say, 1996 can be determined.

The cheapest alternative in terms of present value can be seen to be Table 8. Total costs of this alternative (including common costs such as Transmission costs which were not needed for determining the cheapest alternative) can now be compared (Table 9) with the total benefits over the economic lifetime of the chosen type of plant in order to determine an internal rate of return.

The internal rate of return is the discount rate at which present value of net cost streams equals the present value of net benefit streams (i.e. the discounted value of the Net benefits column equals zero). It represents the return on the investment, and permits a comparison with the interest the funder could get if the same money was invested elsewhere.

Table 9. Cost—benefit analysis. 40 MW hydro-electric power station

| Year | Costs (£ thousands) | | | | Benefits | |
	Construction costs	Plant costs	Operating costs	Transmission costs	Tariff income	Net benefits
1996	500	—	—	—	—	(500)
1997	1250	—	—	500	—	(1750)
1998	1500	750	—	1000	—	(3250)
1999	750	750	—	500	—	(2000)
2000	—	—	—	—	750	750
To	↓	↓	↓	↓	↓	↓
2010	—	—	—	—	750	750
2011	—	—	—	—	750	750
etc.	↓	↓	↓	↓	↓	↓

It is postulated that the internal rate of return is, say, 20%

In this example, if a lender could get a 12% return without risk by putting his money into a building society compared with the 20% obtainable by investing in this project with all of its inherent risks, he might consider it better to do so.

Risk analysis by sensitivity testing
A risk analysis of the proposed project can be undertaken by

considering the effect of variations on initial best guess assumptions on the internal rate of return.

Variations for power projects could typically include

- a construction cost increase (or decrease)
- an operating cost increase (e.g. fuel for gas turbine)
- tariff income, which is lower because of increased government taxation, or fewer customers or good summers/winters and hence lower demand.

Essentially, the initial best guess calculations are reworked using the extremes of the most likely range of variations, e.g.

- construction costs ranging from −10% to +40%
- fuel costs ranging from −5% to +20%
- a combination of these.

The range of variables can be selected using subjective experience of specialists or by using published data. Bearing in mind at this stage the ballpark nature of the values used in costs and benefit streams, this approach can be adequate initially for many projects to decide on whether or not a funder will allocate funds. It also does not necessarily require a specialist computer program as all the work can be done on a spreadsheet.

Monte Carlo analyses

A more sophisticated approach to risk analysis using Monte Carlo random sampling techniques[3] is available which is a development of the more simple sensitivity testing procedure. Input is in a form which identifies in detail the range of likelihood associated with each key variable.

A typical basic procedure is that a computer model is established for a particular project which randomly selects combinations of sensitivities from the ranges of operator-estimated possibilities applicable to each key cost item (e.g. for the construction costs above, somewhere in the range −10% to +40%). It repeatedly undertakes internal rate of return (IRR) analyses similar to that previously described, possibly up to 1000 times. This can produce a number of alternative presentations of output data, including an IRR curve from which the probability of variations from the ballpark best guess IRR value can be measured. In Fig. 3 the graph shows that whereas the best guess IRR for a project was 36·8%,

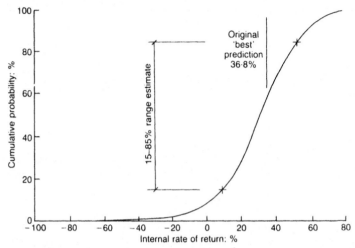

Fig. 3. Output from Monte Carlo analysis [3]

there was a 15% chance that the IRR might be 10% or lower and that there is a 15% chance that the IRR would be 50% or higher. A similar exercise can be undertaken for a range of possible times to complete the critical works on a programme to determine the likelihood of completion in the best guess timescale, i.e. the chances the programme might or might not be met and by how much. Sometimes, computer models can interlink costs and time elements.

However, the initial input best guess assumptions will be no more accurate than those for the more simple sensitivity testing procedure and, furthermore the significance of individual key variables within the overall results can only be obtained if complementary sensitivity testing analysis is also undertaken. The basic process is more geared, therefore, to total project cost or timing risk identification than to risk management by identifying and addressing the impact of changes in individual variables. In addition, whereas a range of probabilities is input for each key variable, those are also usually subjective estimates. Therefore, sophistication of output must not obscure the degree of accuracy relating to the initial assumptions.

Operators of the Monte Carlo method state that a fairly simple study could take two analysts six weeks, plus other staff time for initial input and review. Personnel need experience and suitable risk analysis training, together with computer software and specialist advice. This procedure would be appropriate, therefore, for those funders who require detailed output presentations of a range of likely project costs and timing estimates and have the time and money to do so. Once the computer model is set up, estimates can be updated readily throughout the project as more accurate information becomes available. It is, therefore, a useful project management tool.

MERA cost estimating method

Another method that is used increasingly by UK government departments to justify the allocation of funds is the MERA[4] method. As will be seen, this method has the advantage of being applicable to project estimates at any level of detail, it is simple to calculate manually or by spreadsheet. However, it is not itself time-related in that it does not allow for the spread of costs and benefits over a number of years. It still relies upon subjective assessments of likelihood and does not permit an accurate statistical determination of the chances that calculated values will be exceeded. Stages are as follows.

- A best guess project estimate of known activities is undertaken.
- Possible additional hazardous activities (i.e. those that might result in increased costs) are identified and ways determined to eliminate or reduce them or reduce the possible consequences. Residual hazards are then considered.
- Residual hazards are separated into fixed and variable. A fixed hazard is one which will be needed in total if it occurs, e.g. an additional transmission tower. A variable hazard is one which, if it occurs, may have a range of costs, e.g. possible increases in costs of local materials on an overseas contract.
- For the fixed hazards, the maximum likely cost is determined (this is not necessarily the maximum possible cost). Then the most likely additional cost is determined by multiplying the maximum likely cost by its assessed chance of occurring. A subjective assessment on a scale 0 to 1·0 as described earlier will probably be used; e.g.
  o most likely cost = maximum likely cost × 0·3.
- For the variable hazards, the most likely additional cost is defined as that additional cost that is assessed as having a 50% chance of being correct, whereas the maximum likely cost is

that cost that is assessed as having a 90% chance of being correct; e.g. assuming the best estimate of material costs is £300 000 it may be assessed that
o most likely additional cost = cost assessed with 50% confidence = £50 000
o maximum likely cost = cost assessed with 90% confidence = £150 000.

In other words, the best guess material costs are £300 000, there is a 50% chance that they might increase by another £50 000 and a foreseeable but far more remote chance that the increase might be as much as £150 000.

• All residual hazards are then tabulated with their corresponding most likely and maximum likely additional costs. Whilst the sum of the most likely additional costs is accepted as being a contingency sum to be added to the best guess risk-free estimate, a statistical approach called the root mean square (RMS) method is applied to the maximum likely additional costs in recognition of the fact that all of those are unlikely to occur together at the same time.

The method illustrated in Table 10 is as follows.

Table 10. Example of MERA cost estimating method

Risk	Most likely addn. cost (£)	Maximum likely addn. cost (£)	Difference (£)	(Difference)² × 10⁶
Additional tower	30 000	100 000	70 000	4900
Increased cost of local materials	50 000	150 000	100 000	10 000
	Sum = 80 000			Sum = 14 900
			Square root	£122 000

Best guess risk-free estimate	£500 000
Contingency to cover most likely additional costs	£ 80 000
Best guess including risk allowance	£580 000
Maximum likely additional risk	£122 000
Maximum likely project cost	£702 000

53

- The best guess total project estimate is determined elsewhere as £500 000.
- The most likely additional cost is calculated as £80 000.
- The difference between the most likely and maximum likely costs of each hazard is determined. That value is squared.
- The square root of the sum of those squared values is calculated as £122 000. This is added to the best guess plus the most likely additional contingency value to give a maximum likely overall project cost.

The method can be used directly as indicated above, either for a few principal hazard areas or for a very detailed assessment of every identified residual hazard. It can also be used as part of the simple sensitivity testing method described earlier, for instance to provide data to be used in cost streams.

18 Risk considerations of clients

Risks retained by a client will to a significant extent depend upon what terms exist within contracts between the client and other parties to a project. (The client will also retain other liabilities outside contract pertaining to tort, statute or strict liability as described earlier.) Contractual relationships will vary from contract to contract. One example, shown in Fig. 4, will be used for the purposes of discussion later in the text.

The choice of type of contract (e.g. schedule of rates, design and build), the appropriate standard form (e.g. Institution of Civil Engineers (ICE) 6th, New Engineering Contract (NEC)), and any amendments made to them, can affect the allocation of risks between the parties. The client can select, therefore, the contract and terms which most effectively allocates contractual risk in the way which best represents its interests.

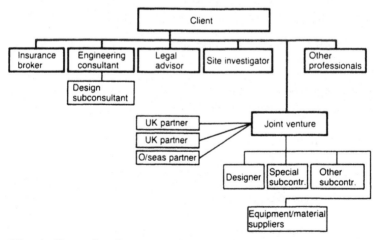

Fig. 4. Example of contractual relationships

Typical construction-related risks which may result in delays and/or increased costs to a client

A detailed assessment should be made of all the hazards associated with a particular project. The following lists cover some of the more usual risks. The reader is reminded that some general commercial risks have been listed earlier in the text and that many of those (e.g. overseas and political risks), although not repeated here, will be applicable equally to construction clients. Particular construction risks for clients can include

Third-party-controlled risks

● Approvals
  ○ planning approvals
  ○ use of hazardous substances on site consents
  ○ tree preservation orders
  ○ conservation area consents
  ○ scheduled monument consents
  ○ need for environmental impact assessments

● Public inquiries
● Legal agreements
  ○ rights of way
  ○ rights of light
  ○ wayleaves
  ○ noise control requirements
  ○ sites of special scientific interest

● Pressure groups, local protests
● Industrial action
● Terrorism in some locations, particularly overseas.
● Changes in regulations
● Changes in statutory legislation

Inherent site-specific risks

● Access restrictions or limitations
● Existing occupiers/users
  ○ alternative provisions (e.g. accommodation, parking)
  ○ working hour restrictions
  ○ maintenance of access (roads, footpaths, associated barriers)
  ○ maintenance of services

- Effect of existing buildings
  - need for protection
  - need for demolition

- Existing site boundaries
  - need to protect and keep in good condition
  - need for temporary enhancement to suit works

- Security
  - protection of works in hand, including temporary works, particularly if several contractors on the same site
  - security provisions including staffing

- Additional land requirements
  - for permanent works
  - for temporary works or for access

- Requirements for new services from statutory undertakings
  - these will include water, gas, sewage, electricity, telephone

- Use of existing services
  - their availability to suit site requirements, capacity, conditions
  - need to determine location, effect of disruption
  - need for relocation around site

- Known ground conditions
  - extent of pre-construction investigations
  - soil types and variability
  - possibility of mining works, subsidence
  - contaminated land

- Climate and weather conditions.

Direct client-controlled risks

- Inaccurate or insufficient terms of reference
- Changes in requirements
  - occupancy, usage, size, scope pre- and post-contract award

- Changes to timescale
  - late decision taking
  - late handing over of site
  - postponements, accelerations or delayed programme
  - early handover of part or whole of contract

- Financial implications
    - availability of funds, generally
    - availability of funds to meet payments due to contractor or needed for third parties
    - liabilities to others if contract completed late

Design-team risks

- Inaccurate interpretation of terms of reference
- Errors in design, contract documents, drawings
- Failure to meet required timescale
    - for producing various phases of duties as required by client
    - for co-ordination of subconsultants
    - delivery of drawings and information to contractor

- Estimating inadequacies
    - changes in labour, plant, material costs
    - inflation
    - taxation changes

- Experience of team
    - experience of members of team
    - continuity of staff

- Type of design
    - established design methods or prototype/leading edge/ unusual structure type designs

- Inadequacy of client protection
    - professional indemnity insurance, collateral warranties
    - provisions for maintaining foregoing after handing over works

- Liquidation/insolvency of members of design team.

Contractor risks

- Failure to meet programme
    - inadequate resources, estimates of duration of activities
    - poor co-ordination of subcontractors
    - inclement weather

- Price changes permitted under certain contracts
- Disputes and claims

- Poor workmanship
- Poor site management, quality control, staff experience, continuity
- Accidents or injuries for which client may retain responsibility
  - under contract
  - due to client staff or joint site occupancy
- Latent defects
- Liquidation/insolvency of contractor.

Other risks

- Changes in conditions affecting viability of completed project
- Political change, i.e. those leading to changes of project scope or cancellations, sanctions and embargoes, tighter exchange controls, repatriation of funds
- Government legislation
- Uninsurable risks, e.g. *force majeure.*

Identification of technical risks
A technical risk assessment will assist highlight hazardous areas of a project, including those for which a client must make provision.

Normal risks

There are risks normally associated with the construction of large civil engineering works including those relating to guarding, watching, site safety, poor materials, poor workmanship, poor plant operation and maintenance, theft, arson, poor setting out, etc. These matters are generally at a contractor's risk: an indemnity and/or insurance will usually be required by the construction contract for damage or loss to works, plant, equipment, third party injury and property damage or loss. Advice may be needed from insurance advisors as to the insurance requirements necessary to cover risks which might result both from within and from outside the contract.

Particular risks

There are risks associated with designing and constructing works of any one particular type (e.g. bridges, tunnels, roads, marine works). In addition, particular locations will have their own risks.

Example of preliminary advice prepared for client on particular technical risks associated with an immersed tube tunnel

The following are some of the technical risks that might be associated with an immersed tube tunnel.

Casting basin

- Often a temporary casting basin will be created on an existing foreshore. This may be adjacent to a public roadway, building or other structure. Basin excavation and/or dewatering and/or loading may lead to deformation of the existing structure with a requirement for remedial works.
- The dewatering of a casting basin on what is often permeable ground may result in lowering of local groundwater levels and may affect groundwater abstraction rates more remote from the site.

Transporting tunnel elements to site

- Damage to (and from) river traffic.
- Damage to riverside property.
- Danger of sinking or running aground, obstructing the passage of river traffic.

Forming trench for tunnel elements

- Deposition of silt from excavation elsewhere in river and forming obstruction.
- Damage to existing services on or under the river bed while forming trench.
- Damage to river traffic or third party property while forming trench.
- The possibility of unexpected ground conditions, trench excavation problems, siltation prior to placing units, trench erosion, etc.

Placing of units

- There are similar third party risks to those listed under transporting, e.g. damage to river traffic and property, obstructions.

Susceptibility to damage after placing units

- Apart from workmanship/design problems, assumed to rest

with the contractor, there is the possibility of damage from ships' anchors, ship impact or discharged debris.

Reliance on third parties

- Significant power is required for site construction purposes including dewatering of the casting basin, and thereafter. A reliable supply, public or otherwise, is therefore essential.
- An important feature of the successful operation of a tunnel will be the telecommunication connection between site and a control room. Delays in completion could delay the opening of the tunnel.
- Where one or both shores are owned by third parties, all necessary permits must be obtained in sufficient time for the works to be carried out without unforeseen hindrance.
- A client will also depend upon the co-operation of the local navigational authority to enable works to proceed without unforeseen hindrance to the contractor.

Client risk control

Having identified the general and more specific project risks by risk assessment, a client can consider tackling them by methods including choice of type of contract, e.g. design/build or remeasure, by choice of standard form of contract, e.g. Joint Contracts Tribunal (JCT) or ICE, and by amendments to contract forms.

Where residual risks remain, these may be reduced by guarantees, warranties, indemnities, bonds; insurances and adequate self-funding of residual risks.

Choice of type of contract

Each type of contract will allocate risks differently between the parties. It is possible, therefore, to select one which is more ideally suited to a client's requirements and, if necessary, to amend it to be more effective.

In general terms, Fig. 5 indicates the apportionment of financial risk between client and contractor in different types of contracts.

It seems accepted that generally public sector and similar clients (for example, Housing Associations) will usually tend to prefer types of contract where there is more certainty of price and as many risks as possible are transferred elsewhere. They are thus prepared

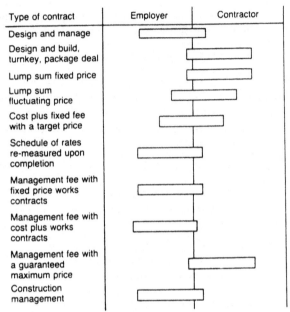

Type of contract	Employer	Contractor
Design and manage		
Design and build, turnkey, package deal		
Lump sum fixed price		
Lump sum fluctuating price		
Cost plus fixed fee with a target price		
Schedule of rates re-measured upon completion		
Management fee with fixed price works contracts		
Management fee with cost plus works contracts		
Management fee with a guaranteed maximum price		
Construction management		

Fig. 5. Apportionment of financial risk in different types of contract[5]

to accept increased tender prices in order to obtain certainty. It will be seen from Fig. 5 that design and build or lump sum types of contract will tend to meet such client's requirements.

Private sector clients on the other hand are more likely to accept more risks because of the opportunity of obtaining both reduced tender prices and the chance of gain if their own wider retained risks do not occur. It will be seen from Fig. 5 that a remeasurement contract, for example, might be selected for that purpose.

Experience or advice from advisors may also indicate to a client that certain types of contract may in practice incur more inherent quality risks, e.g. those with less client-controlled supervision may be more prone to latent defects or higher maintenance costs than those with full client site supervision. For example, for design and build projects there is often little or no independent supervision of the works on site. The result may be a within specification, albeit lower, quality of finished product with higher long-term annual maintenance costs. Where outside funding relates primarily to capital expenditure, a client will be liable for the annual maintenance costs and matters such as this will be of substantial importance.

One method of reducing such a perception of risk might be to use a high quality specification, a design to above minimum British Standard requirements and independent site monitoring.

Risk bias of specific different standard construction-related contract forms

There are a significant number of alternative standard construction contract forms used in the UK. They are for different purposes, e.g. civil engineering, building works, electrical/mechanical works; they include for different ways of reimbursing a contractor for works undertaken, e.g. fixed price or remeasurement; they allocate risks between the parties in different ways; they may be prepared jointly by a certain sector of the construction industry or may represent the requirements of only one party in that sector, e.g. a client. The following are some of the more common standard forms.

JCT building contract forms

The Joint Contracts Tribunal family of forms cover most aspects of building (as opposed to civil engineering) works. They were negotiated between the various elements of the building industry with the intention of creating a fair balance of risk and reward between the parties. They are specifically designed for building works and do not, for instance, handle ground risks with the sophistication required for the more uncertain civil engineering projects.

ACA form of building contract

This form was produced in 1982 to provide a clearer and more flexible alternative to the JCT form. It was produced unilaterally by the Association of Consulting Architects (ACA) and has been partially adopted by the British Property Federation (BPF) for use by them. It is seen to represent predominately the interests of the promoter.

GC/Works/1 and 2

These are forms of contract developed by the UK Property Services Agency for use by central government on building and civil engineering projects. They are contracts drafted by an employer and substantially designed to protect the employer's interests.

ICE Conditions of Contract

These are produced jointly by ICE, Association of Consulting Engineers (ACE) and Federation of Civil Engineering Contractors (FCEC). They have a variety of provisions for payment of extras such as unforeseen ground conditions, delays introduced by the employer/engineer, and for any lack of certainty within the contract documentation. Accordingly, it is a form of contract which is suited for claims engineering by a contractor. The final contract sum will often be considerably higher than the tender price. There are fairly rigorous clauses requiring a contractor to complete on time, and to carry out the work expeditiously. There are also provisions for extensions of time for a variety of reasons.

New Engineering Contract

The NEC is solely an ICE sponsored document. In fact it is a series of documents covering most common types of contract, ranging from remeasurement to fixed price. It has been suggested[6] that up to a third of government funded contracts should use the NEC by the year 1998. This may be optimistic, but even if only partly met, the NEC is about to become a very significant standard form of contract. The main objectives of the NEC are stated as being flexibility, clarity and a stimulus to good management. To reduce confrontational situations, various precise procedures are specified to obtain co-operation between a contractor and an employer. At the time of writing, the effectiveness of the NEC in achieving its proposed aims is yet to be seriously tested by parties to a contract or, where appropriate, by the courts.

ICE design and build

Payment is a lump sum with a limited opportunity for claiming additional payment. If the employer does not change requirements and the employer's representative deals with matters entrusted to him in a timely manner, virtually the only opportunity for a contractor to claim extra is if unforeseen ground conditions are encountered. Carefully drafted employer's requirements can restrict the opportunities for claiming extra even under this heading. The form is more appropriate, therefore, where employer's requirements do not change and where achieving cost and time targets is paramount.

Contractual methods of controlling residual client risks

Parent company guarantees

Such guarantees are useful if there is any doubt as to the continuing existence of a subsidiary contracting company either during the duration of a project or after completion up to the end of any limitation period. Risks arise because in those circumstances the subsidiary company is not available to provide continued technical services or to be enjoined in any dispute. The guaranteed existence of the parent company may also be a potential risk. In the case of individual professionals or companies which may not be part of a larger organisation, parent company guarantees cannot be provided. The wording of parent company guarantees provided by the individual partner companies for the purposes of a joint venture may be worded to expire when the joint venture ceases to exist, perhaps on the date of the final completion certificate. The exact wording of client-required parent company guarantees must be determined with the advice of the client's legal advisors to adequately reflect the client's concerns and interests.

Collateral warranties

Collateral warranties can be used to create a contractual relationship between parties which hitherto did not have a contract. A contractual relationship can be established, therefore, between a client and a subcontractor or subconsultant. This enables a client to pursue them in contract after a project has been completed and if latent defects occur. Such warranties may be more important where local territory law does not permit ready recovery for economic loss outside of contract due to negligence. Advice should be obtained on this.

Bonds

A bond is normally provided by an insurance company, bank, or specialist surety company. There are different types of bonds that can be used to reduce risk.

- A bid bond assists to ensure a contractor will stand by his tender bid.
- A performance bond can ensure that if a contractor defaults, money is available so the project will be completed in accordance with the terms of the contract.

● A labour and material payment bond protects a client against unpaid third parties reclaiming possession of materials on site.

It is essential that an appropriate value of bond is obtained by a client for each situation and that it fully covers the period at risk. At the time of writing, the wording of typical construction bonds is under review.[7]

Contract wording

This can be used to define the allocation of risk between the parties. In the UK such wording must comply with the Unfair Contract Terms Act.

Retention

Client risk can be ameliorated by the amount of retention agreed to be withheld from the contractor during construction.

Indemnities

These may only be of practical value if backed up by adequate risk financing provisions or insurance. A client will commonly require a contractor and others to provide indemnities. An indemnity is a promise to accept liabilities which may be incurred, e.g. to make good damaged property to the same standard it was before damage occurred, or to pay monetary damages equal to a court award if injury or death occurs.

Insurance

A client may require insurance cover for the costs of its own risks, or cover by professionals with whom it is in contract, or by others such as contractors usually as a back up to indemnities.

If being relied upon, insurance must as far as possible cover the risks for which the insured is liable, the extent of cover must be adequate (and remain adequate) in terms of insured value per event and in aggregate, and the insured must have adequate resources to cover those losses which are not or not wholly insured.

Contractor's insurances will normally include insurance of the works against fortuitous loss, and insurances against liabilities for death or injury and loss or damage to property. A contract will specify which insurances are obligatory and will require a contractor to provide proof of cover.

Limitations to all insurances, including contractor's and professional firm's insurances, can include

- insurance cover which may not fully coincide with the requirements of the client
- insurance which accidentally or purposely is permitted to lapse or the sum reduced
- a sum insured which may not be large enough, coupled with insufficient funds to cover non-insured liabilities
- cover currently made available by insurers being withdrawn (e.g. fitness for purpose not being available in future years when a claim is made)
- claims not met, or met fully, due to a ceiling on annual insured indemnity and other claims having been made on this or other contracts reducing the cover remaining available
- insurance not covering non-negligent acts, or criminal costs resulting in fines
- the insured ceasing trading
- excesses which the insured will agree will not be covered by insurance.

Typical residual risks which may remain with a client

Notwithstanding the general type of construction contract selected, or the standard form or its amendments, or the non-standard contract, the following are typical of some of the project risks which may remain to be considered by a client.

- Excepted risks as defined in a contract.
- Exceptional adverse weather conditions.
- Adverse physical conditions.
- Variations to the contract arising from
  - o inadequacies in the contract document
  - o client requirements changing after award
  - o default by third parties which affects the contract.

- Contract price fluctuations.
- Breach of contract, default or abandonment by a contractor or by a joint venture, or by an individual member company of a joint venture.
- Default or negligence by any of a client's advisors which could include engineering advisors.

- Default or negligence on the part of a contractor's designer, should that contractor cease trading.
- Default or provision of poor material or services on the part of a subcontractor or supplier should that contractor cease trading.
- Latent defects in the works.
- Frustration of the contract.
- Inability of those providing indemnities to meet their financial obligations.
- Inadequate insurance cover.
- The unrecoverable costs of litigation or arbitration.

Some of the above project risks are defined specifically in contract conditions which deal with the entitlement of a contractor to additional costs and extensions of time, subject to various criteria being met. Some risks may be eliminated or reduced by the contractual methods already described. Where non-insurable risks remain with a client, that client should be aware of any limitations on the availability of its own finance to meet non-insured costs and the cost implications to itself of any late completion of the works. Some of these matters are now discussed in more detail.

Excepted risks

A contract will specify excepted risks which do not rest with a contractor. Those risks therefore remain with the client. A client must recognize that not all of its risks are insurable, or insurable economically, and that some may need to be retained, if necessary being covered by some method of risk financing other than insurance.

Excepted risks related to the permanent and temporary works include loss or damage due to use or occupation of the permanent works by the client or authorized persons other than the contractor; problems related to designs which were not produced by the contractor for which a client should require its designers to have professional indemnity insurance; risk of war, invasion, hostilities, etc.; radioactivity; aircraft pressure waves, for which insurance is not usually available and for which a client may need some self-financing in place such as a contingency fund.

Excepted risks related to third party liability for damage to persons and property include crops on site; use or occupation of the land by the client, unavoidable interference with rights of way, or easements, the right of the client to construct the works where they are constructed; unavoidable damage in order to meet the client's requirements; death or injury as a result of a negligent act or breach of statutory duty by the client. Some of these risks are insurable.

Exceptional adverse weather conditions

These can result in delay and normally give an entitlement to an extension of time. What constitutes 'exceptional' is often a matter of dispute and risks can be reduced if the word is more clearly defined. The risks associated with, for instance, high tides, heavy winds etc., can be reduced by appropriate additional wording in the conditions of contract which can define the limits at which risk devolves on to the client.

Unforeseen ground conditions: 'Clause 12'

By definition, the implications of anything unforseen cannot be ascertained. In theory, risks can extend from something relatively insignificant to an issue that threatens the viability of construction and/or involves the client in significant additional costs and delays. Risks can be reduced by a detailed site investigation.

Risk of variations to the contract

Recovery of additional costs which arise from variations required due to inadequacies in contract documentation can be covered sometimes by the client's contractual relationship with the advisors responsible for preparing the documents. A client should have a direct contractual agreement with its principal advisors and collateral warranties with subconsultants. Advisors and subconsultants should be required to provide an indemnity backed up as far as is reasonable by professional indemnity (PI) insurance, the characteristics of this having been defined by the client. In the event of inadequacies due to not complying with specific contract requirements or due to negligence on the part of the advisors, the client can sue them.

If a client's requirements change after award of the contract the cost of the changes would be borne by the client. Such changes must therefore be kept to a minimum.

The risk of default by a third party to the construction contract could be reduced by a formal agreement with the client comprehensively detailing agreed requirements, duties and obligations.

Contract price fluctuations

A clause is usually included in contracts detailing how price will be valued. The risk of price fluctuation may therefore by borne by the client unless excluded by the contract.

Materials on site

Provision may be included in a contract for payment for materials on site. Should payment be made by a client to a main contractor but not passed onto a supplier before that contractor becomes bankrupt, in certain circumstances the client may not be able to prevent a supplier removing its unpaid-for materials for which it retains title from site. A solution is to exclude from a contract any clauses providing for client payment for materials delivered but not incorporated within the works.

Breach, default or abandonment of contract by contractor

A breach of contract is not uncommon but it need not disrupt the progress of the works. A usual approach to encourage performance is to withhold money.

If the contractor excessively delays beginning, or abandons the contract, the client has recourse to compensation from retentions held or if a bond is in place. Risks associated with the failure of an individual member of a joint venture can be reduced by requiring parent company guarantees to be provided by each member for the period of risk.

However, before compensation is obtained a client may be liable in the short-term for the costs of restarting the project with another contractor, for delays, and for any increased prices from a new contractor.

Default or negligence by a client's advisors

Legal advisor. A client's legal advisors will normally have been appointed directly by the client and have professional indemnity (PI) insurance as individuals or as a firm. Risks relate to the legal services provided, including responsibilities for advising on appropriate conditions of contract, warranties, novations, bonds, parent company guarantees, etc.

Client's designer. A client's designer should have an appointment directly with the client and have PI insurance. Risks relate to technical services provided for by the design contract agreement, including specification preparation, supervision (if any), etc.

Client's designers' subconsultants. The subconsultants should be under appointment directly with the client's designers, and have PI insurance. Risks are as for the designer. Should the designer cease trading, there is no contractual link between the subconsultant and the client unless a collateral warranty is in place creating such a link.

Site investigation company. Its appointment would usually be directly with a client. Risks relate to the adequacy of their investigations. Liability relates to whether their investigation was limited in scope by the client or others, was a factual report to be interpreted by others (e.g. the client, their professional advisors, the contractor, etc.) or if a factual and interpretative report was specified. It should be clarified what insurance cover the company has, particularly when operating in a contracting and advisory role.

Insurance advisors. Risks relate to the adequacy of the insurance cover and/or alternative risk financing proposals and risk management arrangements (if any) for which advice is provided. Professional indemnity cover should be investigated.

Professional indemnity insurance. This ensures that professional firms will have funds in the event of a successful claim against them. It usually covers negligent acts and omissions in the work undertaken by them, which they are usually obliged to undertake with skill, care and, sometimes, diligence. Unless they are purporting to be experts in their field (in which case they might be assessed to have a higher duty), consultants have a duty corresponding to what would be normal for professional firms of similar standing in their field: they are not infallible, however, and can make mistakes without necessarily being negligent. There must also be damage suffered as a result of the breach of duty for a claim to be successful. If matters should go to litigation to determine if 'negligence' (a legal term) has taken place, there is always a risk to a client of no or only partial recovery. There are other unreclaimable costs such as management time spent by a client as plaintiff in pursuing a claim. Other limitations associated with insurances have been described earlier.

71

Default by contractor's designers

Contractor or joint venture designers (who may include civil engineering and electrical and mechanical engineering designers, and perhaps other professionals) would be under direct contract with the contractor or joint venture but would have no contract with a client. As far as a client is concerned, any contractor or joint venture design-related problems should be a matter for them. Should a contractor or joint venture cease to exist and design-related defects are found, besides there being no direct contract with a client, the client might find that the obligations entered into by the designers are more limited than those agreed between the contractor or joint venture and the client.

The risk of default or negligence on the part of the designers can be dealt with by requiring them to sign a collateral warranty to establish a contractual link with the client.

Default by subcontractor or suppliers

The problems are similar to those of the contractor's or joint venture's designers. Default on the part of a subcontractor or supplier can be dealt with using collateral warranties or appropriate indemnities from the main contractor. The law relating to the supply to goods for the territory of the project may also be usefully investigated.

Latent defects

Latent defects in certain types of construction, presently mainly building works, may be covered by insurance. However, the cost of such insurance is high and is only currently available from a very limited number of insurers. Costs of premiums will, of course, be ultimately borne by clients as part of a tender price. With properly supervised works the risk of latent defects occurring will be reduced. At the time of writing there are proposals to make latent defects insurance a compulsory requirement of building contracts.[6] However, there is no legislation to compel insurers to provide cover to all contractors (for example, to those with a poor claims record or with little relevant experience or for particularly innovative structures with possible long-term problems) and this may well result in reducing numbers of available tenderers with a consequent rise in prices.

Should a latent defect type insurance not be available or too expensive, a client will need to depend upon having sufficient funds to rectify any ongoing defects if any should appear, and rely upon litigation to recover costs where any party has been negligent or in breach of contract. Some of the risks associated with the latter have already been discussed. Latent defects and their methods of risk management will clearly need some detailed consideration and prudent decision making by clients.

Late design changes or approvals

Design/construct contracts are particularly vulnerable to claims for increased costs due to late design changes or decisions. These must therefore be minimized whether by the client, funders or other third parties.

Late local authority approvals may constitute a problem for a client which might delay the start. There may be problems also in obtaining wayleaves or dealing with local utilities.

Frustration of the contract

The possible causes of frustration of a contract are so wide ranging that risk management strategies cannot be usefully included in the present text.

Costs associated with litigation or arbitration

There are a number of risks should a dispute proceed to a court or arbitration hearing. There is the cost of uncertainty of winning which may lead to settlement at significantly less than is initially claimed; administrative costs; legal and expert costs (both paid before a settlement is reached and before monies received); sometimes unrecoverable defect investigation costs; the unavailability of claimed money until a settlement or legal decision is reached and the money is actually received; the lost opportunity costs of staff who could be more effectively used elsewhere, etc. Litigation or arbitration, even if successful, will not therefore lead necessarily to the recovery of all incurred costs.

Difficulties in pursuing foreign contractors or joint venture partners in litigation or arbitration

The potential difficulties and costs associated with pursuing foreign contractors or joint venture partners in court or arbitration in countries outside the UK are self-evident. International litigation/

arbitration can be lengthy and hence a claimant may need to carry these costs for several years with the added uncertainty of the amount of recovery. Advice will need to be sought from legal advisors as to the form of contracts, choice of law and jurisdiction and so on that could reduce such problems.

Conclusions

A major construction contract exposes a client to a complex set of risks, calling for a wide range of risk management procedures.

It is impossible to eliminate all risks, although many can be significantly reduced.

A client should ensure, therefore, availability of funds to meet retained risks as and when those occur. Methods of risk financing other than by using the direct insurance market have been discussed earlier. All-in project insurance might simplify risk management and this should be investigated.

19 Risk considerations of contractors

The need for a detailed risk assessment for each project has already been described with reference to clients. It applies equally to contractors. The following lists some of the more usual contractor risks. The reader is also referred to other lists included in this text for items which may not necessarily be repeated here.

Typical hazards which may result in delays and/or increased costs to a contractor

Client risks

- Client cancels project
- Client delays start of project
- Client suspends works
- Client delays payments of certificates and claims
- Client delays taking over works
- Client insolvency leaves outstanding debts for work done
- Client insolvency gives receiver material on site for which suppliers have not been paid
- Deficiencies, errors, contradictions, ambiguities in contract documents, e.g. specifications and drawings
- Inadequate supply, quality, timing of information and drawings by client's designers, architects, etc.
- Unexpectedly onerous requirements by client's supervisors
- Unexpected inadequacy of pre-construction site investigation data in terms of interpretation or recommendations
- Unforeseen ground conditions
- Late project changes in size, scope
- Consequences of delays in interconnecting contracts, e.g. plant.

Supplier/subcontractor risks

- Supplier start delay
- Supplier poor performance, quality of materials, timing, delivery of information
- Subcontractor (nominated or otherwise) start delay
- Subcontractor poor performance, quality of materials, workmanship, design, timing, delivery of information
- Subcontractor insolvency.

Constructional plant risks

- Delay in availability of constructional plant
- Poor performance of constructional plant
- Breakdown of key constructional plant
- Lack of standby plant, spares
- Fall in expected resale values.

Direct contractor risks

- Shortages of experienced staff and labour
- Contractor start delay
- Contractor poor performance, inappropriate materials, workmanship, design, timing, management inefficiency
- Poor performance by any joint venture partners
- Liabilities for injury, damage or interference to persons or third party property on or off the site
- Responsibility to correct defects in the defects liability/ maintenance period
- Long-term responsibilities for latent defects which become evident after the end of a contract
- Strikes, labour disputes
- Consequential losses, e.g. loss of profits arising from incidents
- Accident to key operatives, management staff
- Frustration of contract due to some fortuitous intervening event altering the nature of the contract to the extent that it can no longer be undertaken as intended
- Adverse weather
- Fire, theft, other physical risks
- Significant temporary and permanent works failures during construction due to design, materials or workmanship

- Unforeseen services, obstructions, contamination, ground conditions, water inflows, gases, vertical shafts, mine workings
- Archaeological finds
- Explosive finds
- Malicious damage, terrorism, war, civil commotion
- Contractually defined 'excepted risks' occurring
- Unforeseen events and delays not allowed for in tender.

Financial risks

- Inadequate tender pricing
- Unexpected price escalations not covered/covered by contract clause
- Statutory pay increases, tax increases, etc.
- Devaluation
- Financial constraints on ability to meet payments to others.

Third party risks

- Statutory changes affecting construction, e.g. health and safety requirements
- Failure to obtain planning consents, easements, etc. in time allowed
- Unexpected difficulties as a result of interface with third party utilities and others
- Unexpected difficulties as a result of need to liaise with other contractors on site
- Delays in approvals by engineer, client, local authorities
- Local environmental pressure groups
- Damage to works by third parties.

Overseas risks

- Overseas skill/supervision shortages
- Overseas plant, spares and material shortages
- Local customs
- Import/export difficulties
- Exchange control restrictions
- Unexpected exchange rate movements
- Bureaucratic delays
- Remoteness of site access and/or facilities and/or communications.

Litigation/arbitration risks

- Delay in resolving litigation/arbitration disputes
- Uncertainty of result of disputes
- Unfavourable decisions
- Costs of legal processes.

Identification of technical risks

As for clients, a detailed technical risk assessment of particular projects can be undertaken. The initial risks retained by a contractor will to a significant extent depend upon what terms exist within the contract between the contractor and the client (other than liabilities outside the contract pertaining to tort, statute or strict liability as described earlier).

Lists and studies similar to the foregoing can be used to check what risk controls are already in place or need to be implemented in order to reduce potential consequences to an acceptable level compatible with the policy and self-retention capacity of the contractor concerned.

Contractor risk control

A contractor can tackle identified risks by methods including

- undertaking additional site-related investigations of poorly defined risk areas, e.g. sub-soil conditions, likelihood of flooding
- contractual transfer
- using safer working methods plant and materials
- insurances
- adding a risk premium to its tender
- alternative methods of risk financing.

Figure 6 shows how some of a contractor's risks can be reduced.

Contractual transfer

Contractual transfers have been considered earlier in this text and many of the procedures are equally applicable to contractor's risks.

Use of safe working methods, appropriate plant and materials, trained workforce

A prudent contractor will always seek to use safe procedures on

Risk source	Contract conditions	Tender documents	First payment	Advance payments	Currencies of payments	Payments to subcontractors	Limit to penalties	Fixed price quotes	Insurance (various)	Indemnification	Income guarantees	Taxation conditions	Contingency/risk allowances	Investments
Client cancel	●			●					●					
Client fail to pay	●			●					●					
Client suspends	●													
Force majeure incident	●								●					
Contractor cancels	●			●										
Contractor suspends	●													
Client acceptance delay			●	●										
Client start delay	●			●										
Suppliers start delay	●							●	●	●				
Transport start delay	●								●					
Contractor start delay								●	●					
External start delay	●	●	●						●					
Client duration extension	●			●					●					
Suppliers duration extension	●							●	●	●				
Transport duration extension	●								●					
Contractor duration extension								●	●					
External duration extension	●								●					
Base cost estimation error								●					●	
Escalation estimation error								●						●
Exchange rate escalation error	●				●	●								
Investment income estimation error												●	●	
Investment error												●		
Supplier performance failure								●	●	●				
Contractor performance failure									●					

Fig. 6. Tactics to offset contractors' risks [3]

site. A formal hazard appraisal system is required for most site activities under the management of Health and Safety at Work etc. Act. However, risk assessments can be extended beyond matters concerning health and safety to cover other areas of risk. Table 11 shows a typical form that might be used (and almost certainly improved upon).

Works insurances

Many contractor's risks can be covered by a works insurance.[8] The two main types are

● Contractors all risks (CAR), covering damage to constructional

79

plant and the works in progress in building and civil engineering projects
- Erection all risks (EAR), covering damage during the erection and testing of machinery.

Table 11. Simple preliminary hazard assessment for contractor

Item no.	Operation description	Hazard	Assessed degree of risk	Control measures	Assessed residual risk
5	Excavation of foundations in zone 2	Reduction of support to adjacent third party	3	Trained plant operators Pre-excavation condition survey Defined haul routes and ramps Design assessment of effects of removal of support Robust barriers to limit approach of excavation plant Daily monitoring by visual observation and gauge reading Weekly monitoring by survey equipment (see Method Statement ...)	1

Degree of Risk
Low = 1   Medium = 2   High = 3

PREPARED BY . . . . APPROVED BY . . . .

Signature   . . . . Signature   . . . .

Date   . . . . Date   . . . .

Typical works insurance cover is illustrated in Table 12. Sometimes public liability and other insurances are included as part of the works insurance package.

Only fortuitous loss and damage is covered, not that which is an obvious possible consequence of a contractor's imprudent way

Table 12. *Hazards and typical basic covers in works insurances*

Hazards	Covers in works policies
	Cover provided by contractor's or erection all risks policies
Fire	+
Explosion, chemical	+
Lightning	+
Plane crash	+
Faulty products delivered to site, e.g. faulty material, faulty design and manufacturing	+
Explosion	+
Short circuit	+
Leakage	+
Windstorm	+
Snow, rain, hail	+
Flood inundation	+
Frost, ice	+
Earthquake, landslide, subsidence	+
Deliberate damage, sabotage	+
Strike, riot, civil commotion	−
War	−
Constructional plant	+
Transportation	Transportation insurance
Loss of profits (following property damage)	Loss of profit insurance
Liability	Public liability insurances

of undertaking works. Different policy wordings provide different extents of cover. Insurers sometimes follow the wordings of one or two main insurance market construction insurers, or sometimes use what are called the DE (defect exclusion) clauses, produced by the market for different circumstances. These can range from excluding all loss of, or damage to, insured property due to defective design, specification, materials or workmanship; to all the foregoing exclusions but including cover for other insured property damaged as a consequence.

Depending on the country in which the risk is situated, individual hazards may be included in or excluded from cover by way of the written terms and conditions.

Limitations to insurance cover have been described earlier. To cover those limitations, risk financing alternatives may need to be considered by a contractor as well as the amount of risk that can be retained in-house.

Determining risk premiums to be added to tender

Identified residual risks that a contractor cannot transfer or which are believed to be outside insurance cover should be assessed for significance and an appropriate adjustment made to the tender sum.

Traditionally, cost estimates are based on available data for actual costs on previous contracts and on experience, and thus implicitly include an assessment of the costs of risk within them. The actual tender price is often the best guess cost estimate plus (or minus) a percentage adjustment which takes account of commercial aspects such as the state of the market and how much a contractor wants the work, the opportunities for additional reimbursement under the contract and so on. This approach does not normally include a detailed risk by risk evaluation, not least because of time limitations.

However, in some circumstances and if time permits, a more detailed risk analysis approach may be justifiable, either to individual items or to whole projects.

Example relating to individual rate

A detailed risk assessment of individual rates can be undertaken as illustrated in Table 13.

Table 13. The weather effects on excavation cost estimates for isolated column bases in clay[5]

Weather conditions	Probability	Unit price rate (£)	Probable cost (£)	Time in minutes	Probable time
Very dry	0·10	2·60	0·26	12	1·20
Fairly dry	0·20	3·00	0·60	15	3·00
Wet	0·50	6·00	3·00	25	12·50
Very wet	0·20	8·00	1·60	35	7·00
		Probable cost	5·46	Probable time	23·70

The most probable statistical cost is £5·46 with a probable time of 23·7 minutes.

It is not known if such an exercise is commonly carried out by contractor's (or anyone else) estimators, but it will be appreciated that at the very least it is a neat mathematical solution to such a problem if and when one arises.

Example relating to whole project

In a design and construct road widening scheme the following risk assessment process was described.[9]

- The design and construct team identified as many hazards as possible. Unknowns were reduced as far as possible by further investigation.
- The monetary consequences and likelihood of occurrences was determined for each hazard. Table 14 shows the probability scale used.
- The product of the monetary consequences and the probability gave a value for each risk.
- Significant risks were designed out where feasible.
- The total value of remaining risks was assessed and the decision made on what additional risk premium adjustments should be made to the tender sum.

Table 14. The probability scale used in risk assessment

Likelihood	Likely	Even chance	Possible	Unlikely	Very unlikely	Very very unlikely
Probability	100%	50%	33%	25%	10%	Nil

The foregoing procedures have all been described previously in this text. Table 15 illustrates the resulting risk analysis.

In the event, none of the pre-identified risks occurred but a totally unforeseen hazard did, of which the costs to the contractor were off-set by the £8000 identified by the risk assessment. The latter circumstance is not unusual: the largest insurance loss on the Humber bridge was from the effect of pigeon droppings on warehouse-stored spools of cable wire. Understandably, such an occurrence was not anticipated by the underwriter's risk assessment!

Table 15. Risk analysis for road widening scheme

Hazard	Result	Action	Cost (£)	Proba-bility	Value of risk
Natural Solution features	Soft areas of formation to be replaced with compacted fill	None. Solution features considered unlikely in this location	Nil	Nil	Nil
Very inclement weather	Site standstill	None. Sandy soil will allow quick resump-tion of work	Nil	Nil	Nil
Man-made Archaeo-logical sites	Project delays	Archaeologists consider finds in this area very unlikely. No action	Nil	Nil	Nil
Unexploded ordnance	Site shutdown for 2 days during bomb disposal exercise	Allow 2 days standing on labour and plant	5000	1:10	£500
Methane gas from waste site	Possible escape into drainage system creating explosion risk	None. Cut off trench and verge width make this result too unlikely to consider	Nil	Nil	Nil

Table 15 *(continued)*

Hazard	Result	Action	Cost (£)	Proba- bility	Value of risk
Unexpected services	Project delays	Gas main position known. BT cable positions clear of works. Authorities confirm no water, oil or electricity services. No action	Nil	Nil	Nil
Design Depth of existing carriageway not known	Possible need for additional road base on phase 2	(a) Allow 2500 tonnes @ £30 per tonne (b) Second assessment of records made to confirm design	75 000	1:10	£7500
				Total risk assessed at	£8000

20 Risk considerations of professional firms

General risks

As for all commercial undertakings, professional firms such as consulting engineers, quantity surveyors and architects must ensure that they agree reasonable terms of engagement that will give them an acceptable profit, that they do work in an efficient way to achieve that profit, and that they actually get paid, and paid promptly, for the work they have done. However, one of the biggest risk areas for professional firms is tortious and contractual liability.[10]

Sometimes liabilities are recognized and accepted as commercial risks. Sometimes liabilities are not recognized. It is essential, however, that liability exposures and the cover provided or not provided by professional indemnity insurers are fully understood by those dealing with them, and that proper management procedures are in place for handling such matters.

Liability exposures

For legal liability for damages to result, it must be proven that a professional has a duty to a claimant, that there has been a breach of that duty, and that a legally recoverable loss (e.g. one which is not too remote) has been incurred. Duty may occur in tort, or in contract. For a professional a breach of tortious duty will almost always be one of 'negligence', which is a legally defined term. If a professional discovers an error before construction begins and rectifies that error, there may be no grounds for a client to claim loss as a project may then cost no more than if it had been designed correctly in the first place. (There may of course be consequential losses due, for instance, to later completion because of delays incurred in rectifying the error.) If an error is discovered after construction starts, the professional may become liable for any additional resulting costs incurred by a contractor as a result of the contractor having to reorganize his resources and method of working. The extent of liability will depend also upon the 'foreseeability' of the losses. The extent of foreseeability (or any equivalent legal term) is a matter to be decided by the courts.

The standard of care required by a professional is that of the reasonable average ordinarily competent member of his profession. The professional must be as aware as other competent practitioners of developments in the particular area of expertise in which work is being undertaken. This means that ordinary professionals (more so than those professing to be specialists, or experts, or claiming to use a Quality Assurance management system?) are permitted the occasional error or lapse, and are not expected necessarily to be *au fait* with the more esoteric issues less readily available in the technical press. However, a professional, in practice, will be expected to have up-to-date knowledge of appropriate standards, codes of practice and similar documents.

To cover the consequences of many of the claims being made, the professional will usually be obliged by clients to have professional indemnity insurance. This is sound commercial practice anyway.

However, it is essential that professional firms realize that their conditions of engagement (their agreements) may incorporate onerous terms that are

- outwith professional indemnity insurance cover
- likely to lead to an increased likelihood of insurance claims and hence premium and/or excess increases
- unreasonable and impose responsibility for actions over which they have no control.

Professional indemnity insurance

Professional indemnity insurance is in respect of legal liability arising out of the conduct of the professional during the course of that professional's agreed business. Normal limitations will be claims above a pre-agreed ceiling *per annum* and claims below a pre-agreed excess.

Some of the most significant risks to professionals are those arising out of claims that are outside insurance cover, or which might arise because certain obligations of the insurance contract between the professional and the insurer are not met. The following various aspects of indemnity insurance cover are briefly noted.

Fines and penalties. Insurers will not indemnify an insured against illegal acts or non-compliance with certain statutory requirements (e.g. the Health and Safety at Work etc. Act).

Deliberate acts and omissions. Insurers will not indemnify an insured against the consequences of deliberately risky actions.

Legal liability. Insurers are not concerned with events for which an insured is not legally liable. Legal liability is a matter determined by law and generally proving liability will be a matter for whomever is making the claim against the insured. It will be handled by the insurers on both sides of a dispute, and usually by their legal representatives.

The course of business. Insurers will generally only consider claims arising out of the insured's pre-agreed recognized business. This is not least because insurers will have one scale of rates and terms and conditions for, say, consulting engineers, and another for a firm undertaking contracting activities. Therefore, a consulting engineer should obtain insurer's approval for any proposed work outside what could be considered normal consulting engineering activities. Examples might be the direct hiring of plant; the undertaking of works which, for example, might include physical opening up of a structure for inspection purposes; the direct engagement of others as part of an all-in package to a client, e.g. the complete inspection of a chimney including engagement of steeplejacks and equipment. If hiring of equipment or contracting firms cannot be avoided (e.g. by acting as an agent for the client), risks can be reduced by obtaining indemnities or hold harmless agreements from the suppliers or contractors. However, some might insist on standard industry conditions of engagement which could still have the effect of transferring risk to the professional. There is also substantial case law relating to vicarious liability on the part of those giving instructions to others. As stated earlier, to avoid uninsured exposure, prior approval should always be obtained from insurers.

Reasonable skill and care, fitness for purpose, etc., clauses. The normal duty of a professional in tort and statute (i.e. the Supply of Goods and Services Act) is to exercise 'reasonable skill and care'. This is mirrored in standard professional agreements such as those of the Association of Consulting Engineers.

Under contract that duty can be extended, for instance, to 'all due skill and care' which is more onerous. This would be a contractual legal liability which, as already discussed, would be covered by the professional indemnity insurance. However, being a more onerous duty, there is corresponding higher risk of claims being made. There is thus more direct exposure of a professional to loss because of the deductible costs which would be applicable to each claim. In addition, professional firms with larger claims records can expect higher premiums, larger deductibles and less favourable conditions. Therefore, clauses in agreements which contain duties more onerous than 'reasonable skill and care' should not be accepted if at all possible.

An extremely onerous duty that clients (a term which here can include contractors on design/build contracts) seek to impose on professionals is one of 'fitness for purpose'. This is so wide an exposure that insurers usually exclude it from cover. In soft insurance market conditions it may be available occasionally for certain professionals, but there is never any guarantee that it will be included in future annual renewals, particularly if insurance market conditions improve. Depending on the size, type and value of project (e.g. a contract to design simple estate roads may be deemed not particularly risky), a professional should normally refuse to accept an agreement containing 'fitness for purpose' clauses.

Conditions of insurance contract. It is a condition of insurance that any intimation of a possible claim against an insured is provided to the insurer as soon as reasonably possible. Professional indemnity is provided by an insurer on a claims-made basis — in other words, the insurance cover is for claims made during the period (usually annual) of the period of insurance by that insurer. It will be evident, therefore, that there is a risk of a breach of insurance conditions if there is an unreasonable delay in notification. In addition, at renewal or if a new insurer is contemplated, it is a prime requirement of an insurance contract that all material facts (i.e. those that might affect the judgement of a prudent underwriter) are disclosed — once more, an insurer must know of all potential claims, otherwise cover may be deemed not to exist. A further risk arises if a new insurer is appointed but refuses cover for a claim that should have been made during the previous period of cover by the earlier insurer, whereas the earlier insurer also refuses cover due to unreasonably late notification. The insured then finds itself with no cover.

Conditions which might be included in proposed agreements

Assignment, novation, collateral warranties

Assignment, novation and collateral warranties are contractual methods by which parties attempt to transfer their risks and obligation to others.

- An assignment is a transfer of rights, e.g. the right to sue a professional firm can be transferred from a client to a future tenant.

- A novation is the transfer of rights and obligations, i.e. the obligation of a client to pay fees could be transferred to an unknown third party. It is in effect a new contract between the professional and the third party.

- A collateral warranty is a separate contract between two parties formally not directly in contract, i.e. where no agreement exists between the two parties. However, it may be a requirement of a professional services agreement that a professional must agree to sign a collateral warranty with a third party, e.g. to permit a client to form a direct contract with a contractor's designer. As post contract completion latent damage is in some circumstances more readily recoverable in contract than in tort, this greatly increases exposure to future claims.

From a professional point of view, each of the foregoing should be as far as possible resisted at the agreement negotiation stage. If, for commercial reasons, they cannot be deleted entirely, it is often possible to restrict transfer of, for example, rights to once only, requiring perhaps 'reasonable notice and the agreement of the professional which will not unreasonably be withheld'. After the agreement has been signed, any post-agreement negotiations take place with the professional in a much stronger position. There is no need (except for future commercial considerations) to agree to accept any change of risk allocation at all, or at the very least without some substantial consideration given in exchange. Considerations may be money, or may be a requirement to pay any fees outstanding at that time.

Contracts executed as a deed

The time during which a professional can be sued in normal contract is six years after the completion of a contract, or twelve years for contracts executed as a deed. The latter types of agreements should be resisted, therefore, if possible.

Other issues

It is matters such as the foregoing which should be looked at in all proposed agreements. Other clauses might refer to a requirement to provide parent company guarantees, bonds of various types, liquidated damages, copyright, use of materials complying with British Standards only, guaranteed maintenance of certain levels of indemnity cover for minimum periods, variations on the basic duty of 'reasonable skill and care', use of words such as 'ensure', 'total satisfaction of the client', etc., the taking on of risks outside the control of the professional, and so on.

Clause examples

The following are typical of some of the clauses seen in proposed contractor design/construct agreements with professionals, with some possible interpretations in brackets.

- The designer shall exercise all reasonable skill, care and diligence in the performance of his services. (The word 'all' imposes an increased exposure beyond the normal 'reasonable skill, care and diligence'.)
- The designer shall carry out these structural and civil services in such a manner so as to enable the contractor to fulfil its express and implied obligations under the contract (e.g. fitness for purpose which may be outside professional indemnity cover).
- The designer shall determine any information which may affect the contractor's tender (e.g. be aware of all matters which might affect the contractor's tender price).
- The designer shall advise the contractor on the need for any special conditions of contract and of the unsuitability of any of the goods, materials or plant which are specified within the employer's requirements (e.g. act as the contractor's contract advice specialist and take responsibility for the adequacy of the employer's technical specification).
- The designer shall advise the contractor on a realistic overall contract programme (e.g. be responsible for the contractor's programme and progress).
- The designer shall visit the site as frequently as the designer deems fit to satisfy himself as to the quality (e.g. guarantee the quality of the contractor's works).
- The designer shall provide such design and specification that

will enable the contractor to prepare the most competitive tender and maximize the opportunities to obtain a contract award (e.g. the designer is responsible for considering every option to obtain the most economical design when priced by the contractor — presumably, if a competitor wins with a cheaper option, the designer has not done his job and can have contractor tender preparation costs (and loss of potential profit) charged against him).

● The designer shall provide advice to the contractor on the level of contingency to be included in the tender and also provide an assessment of the construction risk and uncertainty with regard to design development (e.g. the designer is responsible for assessing all construction risks and uncertainty associated with the design and for determining the contingency level to cover those risks and uncertainties).

Clearly some of these clauses are so unreasonable, and more importantly outside professional indemnity insurance cover, as to be valueless even if agreed to. It can only be wondered why they continue to turn up in draft agreements, as at best they can only produce a false sense of security to those proposing them.

It should be obvious that clauses must not only be considered individually but in the context of other clauses and documents forming part of a contract. What might appear to be rather innocuous in isolation might have a totally different risk implication when considered in conjunction with, say, a definition given somewhere else.

Risk reduction procedures

To minimize risks a professional must

● assess all agreement clauses singly and in conjunction with the rest of the clauses for their practical implications in terms of the risks associated with the type and size of project

● consider the firm's financial capacity to meet possible costs outside of insurance cover

● negotiate to delete or reduce the impact of those risks that are of concern

● decide which risks may be effectively passed on to others such as subconsultants with the capacity to meet them

● make a final commercial decision as to whether to sign the final agreement or not.

A proposed agreement may be so convoluted with interconnected clauses drafted in legalese that the only way to approach it is to make in effect a counter offer in a covering letter to the agreement stating something along the lines of 'notwithstanding the clauses in the proposed agreement, the duty of the professional will be no more than to exercise "reasonable skill and care"' or 'no more than those contained within Association of Consulting Engineers Agreement No. 2'.

Management procedures

Contractual liability of professionals is an area of such potential risk that particular measures should be adopted to deal with it.

A first step would be to establish proper communication procedures between company agreement decision makers and those knowledgeable about insurance cover, total company exposures, the possible consequences of claims and so on. It may be that for small practices all such matters are handled at partner level anyway. Alternatively, it may be necessary to ensure that other specialist members of staff are available to provide day-to-day advice to project managers on non-standard agreement matters and to perhaps assist undertake initial negotiations with the proposers of those agreements.

A second step would be to establish standard company procedures for dealing with agreement risks. For instance, risks could be categorized in order of importance. At one end of the scale could be those risks that are outside insurance cover, or otherwise considered impossible to agree by company policy (e.g. parent company guarantees). At the other end could be those risks that are inside cover but the professional would rather not have in the agreement, such as those increasing the standard of care above the 'reasonable skill and care' level. It could then be agreed which risks need to be notified to insurers (usually via an organization's brokers), and which risks needed sanctioning by which levels of the professional organization, taking into account commercial aspects (such as the size and type of job, how much the work is really wanted, maintenance of relationships with the proposers of the agreement, etc.). Typical acceptable alternative form of words could be pre-prepared for the more common problem clause areas to aid negotiation. Brokers, insurers and legal advisors would need initial and then periodic involvement. For individual agreements the final commercial decision to accept risk must be taken by a knowledgeable professional at the level in the firm agreed to be the most appropriate for the size of risk, taking account of all the advice available. As stated previously, these matters might be handled at partner or director level, or individuals may be designated to deal with them. Appropriate training may need to be considered.

It is only by establishing, implementing and auditing the effectiveness of such a strategy that professionals can ensure that this important risk area is under control. The results should in the long-term show that effective control leads to fewer claims and hence better insurance terms and conditions. A direct simple comparison should be possible between the additional salary costs of implementing the strategy and the resulting savings in premium, thus providing an initial assessment of the cost effectiveness of the risk management process.

21 Technical risk considerations of works insurers

Types of contractor's construction insurance policies
Works in progress are usually covered by a contractors all risks or erection all risks policy, as described earlier. There are two main types.[8]

- *Annual or floater insurance*, where terms and conditions are pre-agreed between insurers and a contractor and cover a range of typical projects. The contractor advises the insurer of the value of the additional works gained and pays the increased premium. The insurer is obliged to take on the works at the pre-agreed terms.
- *Project insurance*, where an individual project is sufficiently different, large, or where annual type policy limits are exhausted. Terms are not pre-agreed and the insurer can refuse to provide cover or can require onerous terms. (This is not to be confused with another meaning of project insurance where all of a client's, contractor's and possibly named subcontractor's liabilities are covered by a blanket project policy.)

In the following paragraphs various aspects of construction insurance are discussed and the impact of those on other parties to the construction process.[11,12]

Current insurance risk assessment practice
In the UK construction insurance is mainly broker-led, with construction companies instructing brokers to obtain from insurers insurance on the best possible terms. For project insurance this usually means as quickly as possible (often in a few days) after winning a contract. The technical information made available by

95

the broker to prospective insurers is that provided by the construction company itself. It is for the insurers to assess such information in order to propose terms to the broker. Terms are the product of the underwriter's skills and experience (albeit usually as a non-engineer) in the specialist construction insurance market, and of market forces, and not usually the result of a more rigorous scientific assessment. The more unusual a project, the greater the risk to the underwriter. In recent years insurers have become increasingly averse to riskier projects. While acknowledging that technical risk assessments could help determine more accurately profitable terms and conditions, underwriters find that in practice market limitations (e.g. available time and a natural unwillingness to spend money until an insurance contract is agreed) preclude detailed assessments in a majority of cases.

Consequences for contractors

The net result of conservative underwriting practice is that certain construction risks are sometimes not readily insurable or only insurable by a few insurers whose terms and conditions are higher and more restricted than contractors have been used to expect.

In addition, insurers may exclude more readily specific areas of a main risk from cover; impose larger excesses/deductibles; set onerous conditions precedent to cover being provided; place low ceilings on the amount indemnified; or not agree to include inflation clauses or full reinstatement cover.

Due to the unwillingness of insurers to cover certain hazards on some contracts (ignoring those hazards which are not normally deemed insurable anyway) and the onerous restrictions on cover, contractors are sometimes not obtaining from insurers the insurance cover required by construction contracts. Any shortfall has to be met, therefore, by alternative forms of financing.

Consequences for client

A client may or may not realize the extent of the gaps in insurance cover compared with contractual requirements. However, even if that is the case, a client might consider that whether uninsured or insured, according to the contract an indemnity exists and risks still rest with the contractor. That, of course, is the case for as long as a contractor has the alternative resources to support the indemnities and the contractor remains in business. A major indemnified but significantly uninsured catastrophe can have a significant impact on a contracting business's reserves and hence ability to survive. Therefore, clients may be sometimes more at risk than they realize.

Problems for insurers

Typical problems for many construction underwriters, exacerbased by the lack of expertise in-house for the wide range of projects that might need to be considered, and the extremely limited time scale to do anything other than a brief assessment in order to bid for the work in competition with other insurers, are a lack of specific technical experience of

- the general hazards to works under construction
- the hazards associated with particular types of work, e.g. tunnels or bridges
- the interrelationship between different activities on a site, e.g. layers on a breakwater and/or the changing degree of protection it provides to works under construction behind it
- the possible physical risks, e.g. windstorm, earthquake, flooding, etc.

Also of critical importance in assessing physical risks on exposed sites are

- a knowledge of probability of occurrence of identified hazards during construction periods, e.g. the chance of a 1 in 50 year storm occurring or being exceeded at any time during a three-year contract period
- the significance of having adequate records from which to extrapolate, e.g. risk of damage by flooding to protective temporary works such as cofferdams.

Available information on insured risks

Information used by insurers for assessing construction risks is clearly of value to clients, contractors or others attempting to assess risks for their own purposes. As stated earlier, it must be noted that insurers are only interested in insured risks and not those that fall outside of cover.

Insurers mainly use information from

- broker-supplied information

- insurer's own surveyors
- historical records of similar risks
- reference books
- subjective assessments
- the previous claims record of the contractor
- in-house engineers, only a few (mainly electrical/mechanical engineers) operating in an underwriting capacity.

Only some sources of information are readily and publicly available. A small number of the larger insurance or reinsurance companies maintain their own databases of territory, physical hazards, insured claims that have been made, and methods of transferring this information into useful underwriting criteria, but this is normally not readily available, particularly outside the insurance market.

The Insurance Institute of London publishes a guide[8] available to the public which includes chapters on construction risks. It covers (albeit in general terms) most of the major physical hazards, plus hazards associated with both general and more particular types of construction.

For example, particular risks associated with immersed tube tunnels are stated as

> With immersed tube tunnels, construction problems arise from the size of the individual tunnel units, their construction on land, the preparation of a stable river bed trench, and the subsequent launching, towing and positioning of units on site. These problems will be greater if the tunnel is being constructed in an existing shipping channel. Problems may occur with dewatering casting basins; flooding over cofferdams protecting casting basins or shore works; initial trench excavation may cause siltation elsewhere; the trench itself may silt up prior to or during placing units; tidal/current may cause placing problems; differential bed settlement may cause joining problems.

This description might interestingly be compared with the fuller assessment of an immersed tube tunnel given earlier in the text. It will be seen that the Insurance Institute description provides a useful start to a comprehensive list of possible hazards for those underwriters who may not have been involved in such work previously.

Probability criteria
There is little or no published guidance to insurers or others

in construction as to what weather conditions contractors should design their temporary works for, e.g. against wind, storm, flooding or high tides. This is not normally stated in conditions of contract or British Standards. Insurers, therefore, have no guidance normally as to whether provisions to limit losses are reasonable or otherwise. A common design criteria is the 1 in 10 year event, but this is

- an arbitrary number (e.g. why not choose 1 in 5 or 1 in 12?)
- a number that depends upon having accurate records from which to determine it.

It is usually uneconomical to design temporary works to accommodate criteria appropriate to the full permanent works and thus the temporary works could be at greater risk than the permanent works, but only over a much reduced period of exposure. Table 16 shows the probabilities that might be applicable to temporary works.

Table 16. Probability (in percent) that indicated event will be exceeded during construction period

Average return period of event	Exposure period (years)				
	1	2	3	4	5
1 in 10 years	10	19	27	34	41

This table is based on the formula

$$R = 1 - (1 - 1/T)^L$$

where

R = calculated risk (percent)
T = return period (years)
L = period of exposure (years)

It can be used for parts of a year, e.g. for 15 months $L = 1·25$, and the probability that a 1 in 10 year event will be exceeded is $12·3$ percent.

Many civil engineering works will have a risk exposure period of one year or less. However, the exposure period for other works may be significantly higher, e.g. large earthworks schemes on road projects or a hydro-electric scheme and dam would typically take several years to complete.

Temporary works criteria may be, but are usually not, specified in construction contract documents. Even where information is stated, this is not necessarily binding on an insurer unless the insurance contract also makes it so. However, insurance policies are often silent on this point.

Weather-related risk allocation under the New Engineering Contract

One contract document where relevant criteria are given is the New Engineering Contract (NEC). For example, if certain weather criteria specified in the NEC contract data are exceeded by actual events on site, a potential compensation event has occurred.

The specified weather-related criteria suggested in the NEC for the UK are

- rainfall expressed in mm per calendar month
- the number of days in the month with rainfall greater than 5 mm (indicating the days when outside work might be curtailed by rain)
- the number of days in the month with minimum air temperature less than 0 degrees Celsius (when concreting works might be curtailed), and
- the number of days in the month with snow lying at 0900 hours GMT (in the United Kingdom, Meteorological Office readings are taken at 0900 hours GMT; the time may vary in other countries).

The limiting criteria suggested is the 1 in 10 year event. This is obtained for each month. If the 1 in 10 year event specified is exceeded in any month, and loss or delay can be proved to have occurred, a compensation event, entitling a contractor to additional payment and possibly extra time to complete, has occurred. Otherwise, costs and delays are at the contractor's risk.

In the UK, weather-related data would be obtained for that site from the Meteorological Office. Overseas, similar organizations should be used. Otherwise, reasonable assumptions based on what records are available are required.

NEC suggested weather-related criteria can be amended, deleted, or added to, to suit employer's requirements. For example, the

1 in 10 year requirement can be changed to another frequency such as 1 in 5 years or a different frequency can be applied to different risks, and can be changed from a monthly assessment basis to an annual basis.

If appropriate, the reference to low temperatures, snow or rain can be removed.

Additions can be included to suit the project. For instance, it can be specified that a compensation event has occurred if water level in a waterway alongside the site rises to more than a certain level above datum (e.g. risk of flooding of works, temporary cofferdams, etc.); or if wind speed on site exceeds a certain amount for a certain number of hours or days (e.g. risk to tower cranes); or if weather occurring remote from the site produces damage to the site (e.g. flash flooding).

These matters are clearly contractual and relate to the apportionment of risks between the parties to the contract. However, if a contractor is obliged contractually (but not under his insurance contract) to take account of 1 in 10 year weather events but does not do so for his own economic reasons, and damage occurs for which he is responsible, should insurers pick up the bill? Would insurers decide in such circumstances that the contractor had acted unreasonably by not limiting his losses? A contractor should consider such matters most carefully, therefore, when assessing the risks for which he thinks he is covered by his works insurance.

Importance of adequate records

There is always an uncertainty associated with forecasting events such as weather-related probabilities from a limited number of historical records.

An obvious problem is whether those data represent a number of untypically good years as the calculated risk could then be underestimated. Therefore, it is necessary for a contractor to obtain and use records as up to date as possible. A contractor might be at risk of being considered outside cover if it was believed that he was not taking prudent measures to limit losses, e.g. by not updating design criteria in line with current data. Insurers should therefore check periodically that up-to-date data is being used where an initial data shortage problem has been identified.

There are a considerable number of analytical techniques available for forecasting future events from existing records, some easier

Table 17. Example of how a monsoon season is taken into account when planning the overall project schedule for new marine facilities

Activity	Time: months (monsoon season shaded)			
	Year	1991	1992	1993	1994
	Month	M A M J J A S O N D	J F M A M J J A S O N D	J F M A M J J A S O N D	J F M A M J J
1. Tender period		▬			
2. Bid evaluation		▬			
3. Approval and pre-award negotiation		▬			
4. Contract award		▪			
5. Mobilization and construction of temporary facilities		▬▬▬▬			
6. Reclamation			▬▬▬▬		
7. Dredging works			▬▬▬▬		
8. Jetty legs			▬▬	▬▬	
9. Jetty heads			▬▬		▬▬
10. Mobilize quarry, commence rock production and stockpiling		▬▬▬▬▬▬▬▬▬▬▬▬▬▬▬▬▬▬▬			
11. Groynes, revetments and beach nourishment				▬▬▬▬	▬▬▬▬▬
12. Breakwater			▬▬▬▬▬▬	▬▬▬▬▬▬▬	▬▬▬▬▬

to use than others. Whatever is used should be appropriate for the prevailing circumstances, not least the accuracy of the historical data, without being unnecessarily theoretical. In Appendix 2 a technique called the Gumbel extreme value forecasting method is described briefly.

Weather windows

One way insurers try to reduce exposure is by checking that a contractor is not undertaking particularly hazardous work during historically known seasons of poor weather. An experienced contractor should already have taken such matters into account in the works programme, as illustrated in Table 17. Insurers should also check that adequate technical provision has been made by a contractor to protect temporarily any exposed uncompleted works during poor weather periods.

22 Conclusions

General organizational risk management has been described as well as the particular additional aspects associated with in construction. By now it should be understood that

- comprehensive risk management control by organizations operating in the construction field should include both normal commercial and particular construction hazards
- insurance in one form or another will be an integral part of a risk management programme
- other than where dictated by corporate policy, risk control procedures must demonstrate a potential net advantage to an organization. That advantage is usually measurable in financial terms.

Having recognized these points it is hoped that interested readers will now be able to research further into the technical press with a better perspective and more confident understanding of how the various elements of practical company risk management fit together.

Appendix 1 **Hazard identification techniques**

Introduction
Hazards can only be assessed and managed once they have been identified. A combination of identification methods may be required. Certain methods are more appropriate for some industries and in some circumstances than others.

Techniques
Examples of methods for identifying hazards are

- analysis of available information
- brainstorming sessions
- physical inspections
- check lists
- organizational charts
- flow charts
- HAZOP (hazard and operability) studies
- fault trees — example of HAZAN (hazard analysis) process
- hazard indices, e.g. Dow.

These methods are described briefly in the following text. (The reader is referred to other publications for more detailed explanations.[14,15]) Some of the methods will be seen to be of apparent little relevance to other than perhaps the services elements of construction. However, a description is provided to enable the reader to see what is entailed in, for instance, HAZOP or hazard indices methods, if those are suggested by others for consideration.

Use of available information
As described earlier in the main text, there is not a great deal of published information on construction hazards, although some rests with insurance companies and is not readily available. A

conundrum is that the more obviously potentially hazardous some types of construction are, the more often specialized organizations might be used, or the more care may be taken and hence losses, insured or otherwise, can be less. The active construction risk analyst might find it useful to prepare his or her own lists based on what limited information is published, common sense evaluations and the results of brainstorming sessions of actual projects.

One source of information on hazards and risks may be any records kept by an organization itself on incidents, accidents, unplanned machinery breakdowns, claims made, claims paid from insurances, uninsured losses and so on, in terms of timing (e.g. the periods in which they took place) and location.

Where records are not kept in sufficient detail, one risk management task might be to establish procedures whereby the required information is recorded, collated and ultimately analysed.

To supplement in-house records, other loss-related information may be available from, for example, the Association of British Insurers, trade or professional organizations, government and non-government sources, the organization's own insurance company for company historical statistics, manufacturers' data on plant reliability and schedules of planned maintenance, etc.

Analysis of loss records might enable an assessment to be made of the types, extent, timing, geographical spread of losses, the effects of any introduced risk controls, worsening trends and so on, to be determined. It might indicate areas in which to initiate more detailed investigations.

Brainstorming

Brainstorming sessions are the most unstructured form of the various hazard identification processes but can be invaluable in many circumstances. It is a common method used for assessing possible construction risks.

Essentially, all those who are to be involved in a particular future process, plus those who have been involved in similar processes before, should meet to list all possible hazards, however unlikely.

Those hazards of agreed negligible probability are eliminated (but it must be noted that a large number of small losses can have the same effect as one large one so do not immediately round down such small risks to zero so that they are not properly assessed).

Those hazards that can be eliminated by, for instance, further investigations, or by using alternative materials or methods, are deleted.

Where subjective assessments of financial significance and/or probability are needed, opinions can be solicited and consensus views established.

Physical inspections

Physical inspections are usually undertaken directly or in conjunction with local managers. They are commonly associated with insurers' property surveys or with plant hazard identifications, and there are obvious parallels with, for instance, site safety inspections.

The stages are

- preparation
  - o Look at any previous similar reports, including solicited or unsolicited reports from local managers
  - o Consider the best time of year, the time available and the time necessary to undertake a proper inspection
  - o If possible, in advance, prepare proforma sheets for each area or item to be inspected
  - o If possible, in advance, identify any particular features for detailed inspection
  - o Ascertain personal contacts at site or office, management arrangements, who is responsible for matters relating to risk, insurance, safety, plant, etc.
  - o Arrange meetings with local managers and employees
- inspection
- post-inspection
  - o Implement any actions including safety advice, update insurance valuations
  - o Notifications to insurers
  - o File away report with date for future reference, monitoring and to assist form a basis for subsequent physical inspections.

Checklists

Checklists are commonly used for assessing property and machinery hazards. They are normally prepared by the risk manager, completed by members of staff, countersigned by a local responsible person, and usually have an action column to indicate how and who is to deal with particular problems. There are three basic types of checklist.

Simple checklist/ticklist. Factual: 'Please ensure that all the items on the list have been checked', e.g. fire extinguishers, fire safety notices in place. This type of form only prompts the memory as to what to look for. Little knowledge of risk is required. What the person completing the form considers satisfactory may not satisfy a risk manager.

Yes/no questionnaire. 'Are all fire exit routes clear', or 'Are emergency alarms working?' Yes/No/Action. The need to answer yes or no to each question may be considered to be an advantage over a simple ticklist. The staff member is actually stating that something is the case and where a negative answer is given is also expected to say action has taken place. This places responsibility on the staff member.

Descriptive checklist. Asks what responses to a question are most appropriate, e.g. 'Use of safety boots on site — Occasionally/Most of time/Always?' or 'Effectiveness of planned plant maintenance programme — Good/Bad/Average?' This form takes longer to compile, and there is a need for realistic relevant questions.

Advantages of checklists are that they

- are a reasonable, inexpensive way of generating a great deal of information
- are simple and can be arranged quickly
- permit rapid comparison with earlier years and allow effective monitoring
- are easy to adapt to suit changes or accommodate improvement of ideas
- can enable priorities to be identified.

Disadvantages of checklists are

- they may be completed by others/laymen, therefore possible inaccuracies
- possibility of ambiguities, subjectiveness
- possible delays in receiving forms
- there is no way of knowing how seriously/accurately the form was filled in
- possibility of vested interests altering the accuracy of results.

Organizational charts

Organizational charts permit identification of broad areas of risk. They can be produced showing various aspects of company activity, e.g.

- the arrangement of constituent companies in a group
- the management structure
- the departmental structure within each company.

Typical areas of possible risk that may be identifiable are

- duplications. If there are several areas of similar production, each must be made aware of changes in relevant legislation, codes of practise, thus requiring positive liaison and communication
- dependencies and concentrations, bottlenecks.

Charts can be useful for business interruption planning (e.g. the identification and organization of duplicate sources of supply and production), gaps in the lines of communication, the management of health and safety, security, etc.

Flow charts

Typically, flow charts are used to identify and assess production stages, for instance as raw materials are converted to a finished product. They are used in conjunction with 'what if' tabulations. The process is not quantitative. The most important use is usually to identify the consequences of items of plant malfunctioning and to assess alternative courses of action. For example, if an overseas on-site batching plant breaks down, a flow chart would enable one to look at repercussions on site both up and down the line, e.g. can the delivery of raw materials be halted up the line, is there any spare storage capability, is there a need to completely shut down, what are the effects down line, what contractual commitments cannot be met?

By thinking things through in advance, and noting actions, the basis of an emergency rapid response action plan can be formed. The process involves

- a determination by internal discussion on how a production sequence works

- drawing a production flow chart using
  - rectangle = raw materials
  - circles = processes
  - semi-circle = end product
  - arrow directions
  - insertion of relevant quantities

- undertaking 'what if' questions, i.e. 'What if supply broke down?' 'What if a process broke down?'

The 'what if' process is best structured by a simple table, see Table 18.

Table 18. Structure of a 'what if' process

Stage	Likely events causing loss	Likely causes of event causing loss	Possible consequences
Batching plant	Stops	Fire Explosion No fuel Blockage	Lost production downline Raw material storage capacity exceeded Loss of revenue

A column for future actions can be added. The exercise is done for each stage of production. Broad areas of risk are identified first. Individual perils are considered later.

Advantages are that

- it allows problems to be broken down into manageable stages
- all processes relating to one stage at a time can be seen
- it permits structured thinking of causes and consequences
- it identifies bottlenecks.

Disadvantages are

- significant time is needed for full study
- the chart is general.

Units used are those most helpful for assessment. For example, Fig. 7 represents a day's supply/storage capacity and a percentage split in production from batches.

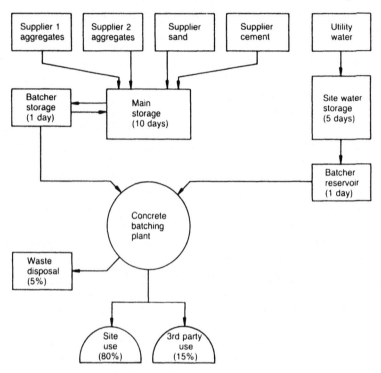

Fig. 7. Simple example of flow chart representing on-site concrete batching plant

HAZOP studies

HAZOP studies are used for the systematic analysis of the individual parts of a process in detail, e.g. the individual parts of a batching plant. The procedure is not quantitative. Analysis is usually carried out by a team who separate the stages of a process into logical constituent parts for detailed study and then identify four main issues, i.e.

- the intentions of the part being studied
- possible deviations from intentions
- possible causes of deviations
- consequences of deviations.

A table of guide words is used. See Table 19. A flow chart is drawn up to check that all parts of system have been covered. See Fig. 8. A detailed analysis is then undertaken. See Table 20.

111

Table 19. Example of guide words, applied to flow of liquid[14,15]

Hazard and operability — guide words		
Guide words	Meanings	Comments
No or Not	This is the complete negation of the intention	No part of the intention is achieved, i.e. there is no flow
More/Less	There is an increase or a decrease in the quantity of the property	There could be more flow than was the intention or less flow
As well as	There is a qualitative increase in the property	The design intentions are achieved but an additional activity occurs, e.g. water gets into the system as well as petrol
Part of	There is a qualitative decrease in the property	Only some of the intention is achieved. This is not a quantitative decrease that would be *less than* but is a decrease in the quality of the property
Reverse	The logical opposite of the intention	Flow is reversed
Other than	The complete substitution of the intention	No part of the original intention is achieved and something entirely different takes place. For example, some other liquid may be put into a system

Advantages are that

- the method enables an extensive identification of possible risks, leading to determination and recording of preventative actions
- it allows each part of complicated system to be examined in detail.

Disadvantages are

- the amount of time involved

Table 20. Consequences of deviation

Guide word	Deviation	Causes	Consequences	Action
No	No flow	1. Tank empty 2. Inlet valve V1 is shut 3. Pump not working 4. Valves V2 or V3 shut 5. Pipe blocked into reservoir	1-5. No water gets to reservoir 4. Water seeps out of pipes 5. Pipes burst	1. Regular checking of tank 2 & 4. Valves to be checked every day 3. Regular maintenance on the pump
More	More flow	1. Pump faulty	1. Spillage from reservoir	1. Regular maintenance
Less	Less flow	1. Pump faulty 2. Valves not fully open 3. Pipe partly blocked	1-3. Longer to fill reservoir	1-3. As for no flow
As well as	Oil as well as	1. Oil in storage tank	1. Oil gets into the reservoir	1. Regular cleaning out of storage tank

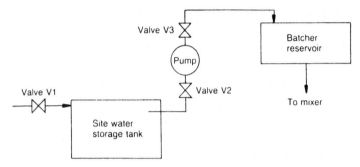

Fig. 8. Flow chart showing water flow into water reservoir

- the need to simplify the system in order to draw a comprehensible diagram and hence parts may be missed out

Fault tree analysis

Fault tree analysis is used to determine potential faults. It is an example of a class of analysis techniques called HAZAN, the systematic analysis of hazards and their potential consequences.

It is used to identify risk in detail while also permitting the risk to be quantified. (Note: this is not a probability tree — see Probability techniques, Appendix 2.)

It permits comparison with

- similar operations within a production area, i.e. is this item more likely to function as effectively as another model?
- required standards, i.e. does it meet minimum probability of failure standards for items of this type?

It is often used to enable weak points in a new device to be identified and tackled at design stage, i.e. it is therefore relevant to new plant design.

Analysis starts with a final event (e.g. a possible explosion or fire) and works backwards to determine possible contributory events. It thus provides a diagrammatic view of how individual events can combine to give a potentially dangerous situation and, by applying individual probabilities, enables a quantification of the overall probability of the final event.

The procedure (see Fig. 9) uses AND or OR logic gates.

- An AND gate is illustrated by a semi-circle shape. Events

114

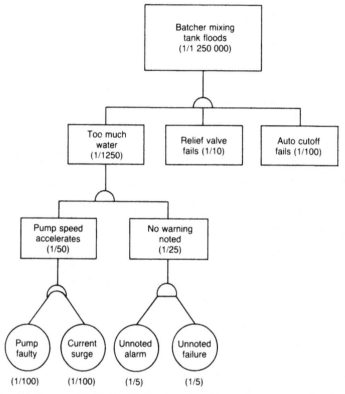

Fig. 9. Events which might lead to batching plant mixer flooding

linked by AND gates should have probabilities multiplied.
- An OR gate is illustrated by a crescent shape. Events linked by OR gates should have probabilities added.

Advantages are that it

- is an excellent structured way of describing a complicated process or system and this aids understanding of that process
- permits risk areas to be identified as tree is being built
- can be used to determine sensitivity of main risk event to changes in parts of the system and permits most effective point of action to be determined, e.g. how increasing the reliability of one part may have a very significant effect on the reliability of the whole process

- permits calculation of the different ways a main risk event can come about.

Disadvantages are

- the time needed
- dependence upon the accuracy of probabilities used which may not be readily available.

Hazard indices

A common hazard index is the Dow Fire and Explosion Index. It is applied to existing process plant by using standard graphs and tables. It enables a comparison to be made between the likely financial consequences of fires or explosions at different locations (e.g. the maximum probable property damage in different areas of a petrochemical installation).

The procedure is that the riskiness in terms of fire or explosion of each process unit in an installation (i.e. the Dow index) is determined from consideration of the type of unit, the types of materials used in the process and the way those materials are utilized (e.g. heat process, pressure process, etc).

Using the Dow index, a radius of exposure is calculated to determine the likely area of damage, and a damage factor applied to represent the percentage damage expected in that area. Mitigating factors such as fire safety precautions are taken into account. Thus, a value can be placed on expected damage due to that process unit.

Similar assessments can be made of all process units in an installation to enable an order of priorities for improvements to be determined.

Appendix 2 Probability techniques

Probability data

Information relating to the probability of organization or project hazards occurring may be obtainable directly or may be derivable from

- historical records kept by the organization
- industry-wide records held by government, non-government, regulatory, trade or professional bodies
- site-specific records of physical risks, e.g. windstorm, earthquake
- manufacturer's data relating to plant
- organization's insurance broker or insurer
- subjective assessments
- mathematical analysis of raw loss data.

Data management and presentation, forecasting, frequency and probability distributions, data descriptions (in terms of mid-point, amount and shape of dispersion, coefficients of variation, skew, standard deviations), combining possibilities and so on, are matters described in all standard statistical textbooks readily available to engineers.

Two particular and practical mathematical aspects are described for ease of reference within this text: probability trees and forecasting from annual events. The former is a standard risk analysis technique that may have some, albeit limited, construction applications. The latter has direct use in determining the likelihood of physical site hazards (such as flooding, windstorms, etc.) occurring, based on existing data.

Probability trees

Probability trees (see Fig. 10) require a logical breakdown of each possible outcome following on from an initial event, and then the

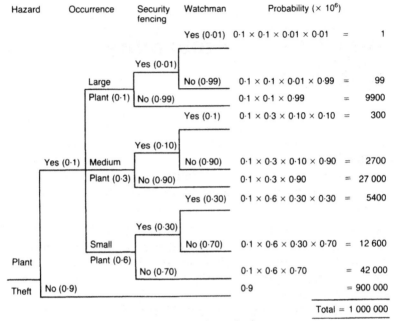

Fig. 10. Probability tree for site plant theft

application of a probability to each outcome. This enables a final probability to be determined for each individual option. By implementing procedures to improve risk at different points identified by the analysis as being critical, risk can be reduced.

Forecasting from annual events

This issue arose in the main text as part of the discussion on the importance to insurers and contractors of the amount of physical hazard data available and the expectations of insurers that contractors will take reasonable measures to limit losses. This they should do by obtaining as much up-to-date data as possible at any one time, and then analysing that data in a reasonable manner.

One of several common techniques for forecasting future events is known as a Gumbel analysis, the detailed explanation for which the reader is referred to specialized texts on the statistics of extremes.

The Gumbel technique takes into account only the single maximum values in each year, i.e. even if there are more than one

118

significantly high value recorded in any one year only the highest one is used in the analysis. For example, if maximum significant recorded flows between 1982 and 1993 were as shown in the second column of Table 21, only those values in the third column would be used.

The procedure is as follows

- the maximum flows are placed in an order of merit, m, the lowest being 1, the second lowest 2 and so on
- the total number of years is R, which in this case is 12
- P, the probability of any flow Q not being exceeded in any one year, is $\dfrac{m}{(R+1)}$; for example, if $m = 3$, $P = 3 \div 13$

Table 21. Example of Gumbel technique

Year	Max. flow (cumecs)	Max. flow used in analysis
1982	133·0 120·7	133·0
1983	74·3 60·0	74·3
1984	145·0	145·0
1985	153·3 110·9	153·3
1986	90·7	90·7
1987	157·0 115·9	157·7
1988	134·4	134·4
1989	106·0	106·0
1990	136·0 121·5	136·0
1991	270·0 240·5	270·0
1992	117·0	117·0
1993	250·0 175·8	250·0

- the value $-\ln(-\ln P)$ is derived. For example, if $m = 3$, $-\ln(-\ln P) = -0.383$

The result can be seen in Table 22. A straight line regression analysis of the points is undertaken. Table 23 shows the criteria obtained.

Table 22. Example of Gumbel technique

m	Q (cumecs)	P $(m/(R+1)$	$-\ln(-\ln P)$
1	74·3	0·077	−0·942
2	90·7	0·154	−0·627
3	106·0	0·231	−0·383
4	117·0	0·308	−0·164
5	133·0	0·385	+0·046
6	134·4	0·462	+0·257
7	136·0	0·538	+0·480
8	145·0	0·615	+0·723
9	153·3	0·692	+1·000
10	157·0	0·769	+1·338
11	250·0	0·846	+1·789
12	270·0	0·923	+2·525

Table 23. Example of Gumbel technique

Q	$-\ln(-\ln P)$
65·97	−1
252·16	+3

The individual points and the line derived from the regression analysis can be plotted on probability paper, as shown in Fig. 11. It will be seen that although there were only 12 years of records, they can be used to derive estimated flows corresponding to any future return period, e.g. a 1 in 10 year flow is about 240 cumecs. Clearly, the more years of data that are available, the more confidence can be placed in the results. It would also seem acceptably conservative to assume that less data might underestimate future maximum flows. It will therefore always be prudent to add some additional value to whatever figure is derived

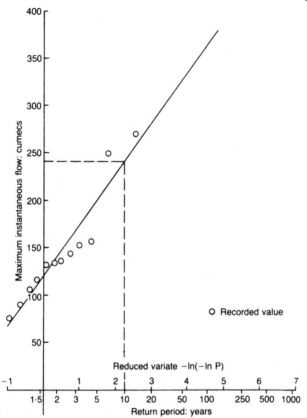

Fig. 11. Example of forecast graph using Gumbel analysis

from this or any similar method, e.g. in the current example, it is suggested that a value of at least 250 cumecs could be justified.

Referring back to how this issue arose in the main text (i.e. that an insurer should be entitled to expect a contractor to make reasonable attempts to limit losses), insurers will hopefully accept a simple approach similar to the above as satisfactory, depending upon the amounts at risk. A contractor is not normally expected to be a statistical expert, records might vary in quantity, and there are various other techniques that might be equally satisfactory, but give slightly different results. Insurers should therefore expect to exercise a degree of flexibility. If the consequences of failure are significant, a contractor might be advised to have independent checks made on his assumptions in order to reduce risks.

References

1. Cassidy D. *Liability exposures*. Witherby & Co. Ltd, London, 1989.
2. Gordan A. *Risk financing*. Witherby & Co. Ltd, London, 1992.
3. Thompson P. A. and Perry J. G. (eds). *Engineering construction risks — a guide to project risk analysis and risk management*. Thomas Telford, London, 1992.
4. Barnes M. Introducing MERA. *Chartered Quantity Surveyor*, 1989, Jan., p19.
5. Flanagan R. and Norman G. *Risk management and construction*. Blackwell Scientific, 1993.
6. Latham M. *Constructing the team*. HMSO, 1994.
7. Edwards L. J. *et al. Construction insurance and bonding*. Institution of Civil Engineers/Thomas Telford, London. Draft of revised edn, 1995.
8. Insurance Institute of London. *Construction and erection insurance*. Draft of revised edn, 1995.
9. Neale D. E. Case study of dualling of A11 Thetford to Bridgham Heath (Norfolk). *Seminar on design and construct-highways*. Institution of Highways and Transportation, North Western branch, 1993, Dec.
10. Taylor P. Minimising the legal risks of professional practice. The professional and the law: preventative measures. *Seminar for construction professionals*. Griffiths & Armour, 1988.
11. Edwards L. J. Aspects of the assessment of technical construction rules by insurers. *Association of Insurance and Risk Managers in Industry and Commerce*, 1992, Dec.
12. Edwards L. J. Building blocks (the use of technical risk assessments to assist control construction risks). *Post* Risk management supplement, 1993, Autumn.
13. Edwards L. J. Weather-related risks under the new engineering contract. *Foresight*, 1994, March.
14. Dickson G. C. A. *Risk analysis*. Witherby & Co. Ltd, London, 1991.
15. British Standards Institute. *Guidelines for the risk analysis of technological systems*. QMS/23, draft, 1993, July.

CPSIA information can be obtained
at www.ICGtesting.com
Printed in the USA
LVOW03s1511211117
557190LV00013B/796/P